The Mystery of Creation

The Mystery of Creation

Adrian Bjornson

Addison Press
Woburn, Massachusetts
www.olduniverse.com

Published by: **Addison Press**
  400 West Cummings Park
  PMB 1725-111
  Woburn, MA 01801
  www.olduniverse.com

Copyright © 2003 by Addison Press

All rights reserved under International and Pan-American Copyright Conventions. No part of this book may be reproduced by any mechanical, photographic, or electronic process, nor may it be stored in a retrieval system, transmitted, or otherwise copied for public or private use, without written permission of Addison Press.

Printed in the United States of America

Publisher's Cataloguing-in-Publication
  (Provided by Quality Books, Inc.)

Bjornson, Adrian
  The mystery of creation / Adrian Bjornson.
  p. cm.
  ISBN 09703231-3-1

  1. Cosmology--Popular works. 2. Creation I. Title.

QB982.B566 2003   523.1
                       QB103-200362

Dedication

This book is dedicated to

Professor Huseyin Yilmaz

*Who has devoted his life to develop his theory of gravity,
which extends the Einstein relativity concepts
to achieve the goal that Albert Einstein sought.*

Acknowledgements

I am grateful for the patient assistance that Prof. Huseyin Yilmaz has provided in explaining his *Theory of Gravity* and the principles of the Einstein *General Theory of Relativity*.

I thank William C. Keel of the University of Alabama for his excellent photograph of the M51 Whirlpool Galaxy with its companion galaxy NGC 5195. This was taken on the 1.1-meter Hall telescope at the Lowell Observatory. I have used this photograph on the front cover and in Figure 1-1.

Foreword

This book is a simplified version of *The Scientific Story of Creation*, by the author, which is shown as Ref. [4] in the Bibliography. Further support for this book is given in Refs. [1] and [3] by the author. The titles of these three references are

Believe [1]:   *A Universe that We Can Believe*
Website [3]:   Internet website, ***www.olduniverse.com***
Story [4]:   *The Scientific Story of Creation*

In this book, these three documents are referred to as *Believe* [1], *Website* [3], and *Story* [4].

This book examines the scientific evidence concerning Creation. Paleontologists have derived a clear picture of the creation of life on earth, since the earth crust cooled 4.3 billion years ago. Astronomers have shown how our earth, our sun, and the stars were created. The book describes these aspects of Creation, which are solidly based on scientific evidence.

As we look outside our Milky Way galaxy, we find that our universe seems to be flying apart. Astronomers explain this by claiming, with absolute confidence, that our universe was created as an extremely dense body that exploded with a Big Bang about 15 billion years ago. Despite this confidence, there are strong scientific reasons to doubt this theory.

After World War II, George Gamow, who had been a leader in developing the atomic bomb, postulated that our universe began with the density of nuclear matter, which he considered to be the greatest possible density of matter. At the Big Bang, our observable universe would have been the size of the Mars orbit. However, Big Bang theorists today insist that the universe began as a *"singularity"* that was *"smaller than a dime"*. A *singularity* ideally means an infinite density of matter.

Singularities have been derived from computer studies of the Einstein General theory of Relativity. Yet Einstein absolutely rejected a *singularity* prediction associated with the black hole that was calculated from his theory in 1939, and in 1945 he rejected the concept that the universe began as a *singularity*. **Einstein insisted that, "singularities do** *not exist in physical reality". After 1939 no scientist claimed that the Einstein theory predicted a singularity while Einstein was alive.*

In 1958, Huseyin Yilmaz published in the prestigious *Physical Review* a gravitational theory that is a refinement of the Einstein theory. It incorporates the principles of the Einstein theory but eliminates its *singularity* predictions. This book gives simple physical descriptions of the Einstein and Yilmaz theories.

In 1948 Fred Hoyle proposed the *Steady-State Universe theory*, which postulated that the universe is infinitely old, and that matter is created to compensate for the universe expansion. This theory had wide support, but fell into disfavor when it could not explain the *cosmic microwave radiation* discovered in 1965. The Yilmaz theory predicts a universe that is similar to the Steady-State Universe theory, but does not have its limitations, and it accurately explains cosmic microwave radiation.

Organization of Book

This book addresses profound issues in a simple manner. To achieve this end, the book has two appendices containing supplementary material, and the book references three documents, described above, which contain supporting information.

The *Website* [3] *www.olduniverse.com*.is available at no cost on the internet. The material in the *Addendum* section (Page 5) of the *Website* [3] is directed toward those with scientific training and assumes the knowledge of calculus.

The documents *Believe* [1] and *Story* [4] are for the general reader, but are more detailed than this book. These documents are described in the internet *Website* [3], which shows how they can be purchased.

If the reader desires more details after reading this book, *The Scientific Story of Creation* [4] is primarily recommended.

At the end of the book is a *Glossary*, which defines terms used in this book, to avoid confusion when unfamiliar words are used. The *Glossary* explains the meaning of exponential numbers like 3.1×10^5 and 3.1×10^{-5}.

The book has two sets of bibliographic references. Those with the prefix Y, such as [Y10], refer to scientific papers on the Yilmaz theory written by Professors Huseyin Yilmaz and Carroll O. Alley.

Contents

Dedication and acknowledgements	v
Foreword	vi
Preface	xii
1. Introduction	**1**
Our mystical Milky Way	*1*
The creation of our sun and our earth	*2*
Life cycles of our sun and other stars	*2*
The universe that lies beyond our milky way	*4*
Meaning of the Hubble expansion	4
Initial size of the Big Bang universe	*6*
Einstein's rejection of the singularity	8
Limitations of the Einstein theory	*9*
The Yilmaz theory of gravity	10
The Yilmaz cosmology model	*10*
The universe predicted by the Yilmaz theory	*11*
Understanding the Einstein and Yilmaz theories	13
2. Creation of the sun, earth, and stars	**14**
The creation of our sun and our earth	14
Early microscopic life	*15*
The evolution of multi-celled plants and animals	*16*
The emergence of humanity	*18*
Was there a Divine spark in the development of humanity?	*20*
The life cycle of the stars	20
The structure of the atom	23
Densities of white dwarf and neutron stars	*23*
3. Our mysterious universe	**26**
The characteristics of our universe	26
Our Milky Way galaxy	*26*
What is a nebula?	*26*
Distances to the stars	*27*
Parallax method for measuring stellar distances	*28*
Measuring the distances of remote stars	*29*
Measuring the radial velocity of a star	*30*
The Hubble expansion of the universe	30
Meaning of the Hubble expansion	*31*
Modern measurements of the Hubble constant	*32*
The observable universe	*33*
The apparent age of the universe	*33*
Alternatives to the Big Bang theory	34
The Hubble expansion is apparent	*34*

The Steady-State Universe theory	35
The Big Bang theory	**36**
Definition of the singularity	36
Cosmic microwave background radiation	38
Spectrum of radiation from an ideal blackbody	39
Blackbody temperature of cosmic microwave radiation	41
Cosmic microwave radiation for the Yilmaz cosmology model	42
The black hole	42
The Schwartzschild solution to General Relativity	44
The black hole prediction	44
Modern acceptance of the black hole singularity	45
The quasar	45
Quasar observations by astronomer Halton Arp	46
Explanations for intrinsic redshift	48
Gravitational redshift	48
The Marmet redshift	49
The mysterious dark matter	49
Understanding the Einstein and Yilmaz theories	50

4. Newton's theory of gravity — 51

The Copernicus revolution in astronomy	51
The astronomical system of Ptolemy	51
The astronomical theory of Copernicus	51
The planetary laws of Kepler	52
Galileo and his telescope	53
The motion of a falling body	54
Newton's laws of mechanics	56
Application of Newton's laws	57
Engineering use of Newton's laws	57
Calculation of the acceleration of gravity	58
The orbits of planets	59
Orbit of the moon	61
Why are astronauts weightless?	62
How Cavendish weighed the earth	63

5. The nature of light — 65

What is a light wave?	65
Mechanical waves	65
Electromagnetic waves	66
Principle of an electromagnetic wave	67
Measuring the speed of light	68
The Michelson-Morley Experiment	70
The contraction hypothesis	71
The Lorentz transformation	72
The Einstein Relativity principle	72
Reaction to the Einstein Relativity theory	73

6. Einstein Special theory of Relativity — 75

Measuring the speed of light	75
The Einstein theory of Relativity	76
The principles of Relativity	77
Explanation of constancy of the speed of light	78
Symmetry of relations between observers A and B	79

x   *The Creation of Our Universe*

Replacing the observer by a set of coordinates	80
Four dimensionality of space and time	80
Variation of mass of an object	81
Proper coordinates for the electron	82
Converting matter into energy	82

7. Einstein General theory of Relativity — 84
- Generalizing the Relativity principle — 84
  - *Equivalence between acceleration and gravity* — 84
  - *Redshift produced by gravity* — 85
  - *Effect of gravity on clock rate* — 87
  - *Effects produced by the sun's gravitational field* — 87
  - *Change of spatial dimensions and speed of light due to gravity* — 88
- Einstein's basis for generalizing Relativity theory — 88
- Verification of general relativity — 90
- The meaning of a tensor — 91
  - *The vector* — 92
  - *The tensor* — 93
- Application of the tensor to general relativity — 96
  - *The metric tensor* — 97
  - *The Ricci curvature tensor* — 98
  - *The energy-momentum tensor* — 98
  - *The Einstein gravitational field equation* — 99
  - *Outline of Einstein theory calculations* — 99
- The Schwartzschild solution of the Einstein theory — 99
- Computer solutions of the Einstein theory — 101

8. The Yilmaz theory of gravity — 102
- Derivation of the Yilmaz solution — 102
  - *The Yilmaz gravitational field equation* — 103
  - *The general time-varying Yilmaz theory* — 103
- Discussion of the Yilmaz theory — 104
- Reason for opposition to the Yilmaz theory — 105
- Consistency with quantum mechanics — 106
- Diagonal character of Yilmaz metric tensor — 106

9. Applying the Einstein and Yilmaz theories — 107
- Relativistic effects produced by gravity — 107
  - *The normalized relativistic mass m* — 108
- Effect of gravity on the speed of light — 109
- The black hole singularity — 110
  - *Misleading astronomical evidence for black holes* — 112
  - *Second limit to the Schwartzschild solution* — 112
- Gravitational effects on distance and clock rate — 113
  - *Effect of gravity on wavelength* — 114

10. Evidence against the Big Bang — 116
- The editorial of Geoffrey Burbidge — 116
- Eric Lerner and Nobel Laureate Hannes Alfven — 117
  - *The Big Bang age dilemma* — 118
  - *Mythological philosophy of Big Bang research* — 119
- The modern Big Bang singularity — 122

11. Weaknesses of the Einstein theory — 123

Cannot achieve a two-body solution	123
Professor Carroll O. Alley	*123*
The single-body Schwartzschild solution	*124*
The analysis of Professor Alley	*125*
Reason for failure to achieve a two-body solution	125
Conservation of matter-plus-energy	126
Multiple solutions from the Einstein theory	126
Variation of speed of light with direction	127
The Einstein theory is not rigorous	127
Finding the truth	*128*
Understanding our Universe	*129*
Einstein's search for a unified field theory	130

12. The Yilmaz cosmology model — 132

Cosmological implications of the Yilmaz theory	132
Description of Yilmaz cosmology model	134
Reduction of speed of light, clock rate, and spatial dimensions	*134*
The Hubble expansion of the universe	*138*
How can gravity make the universe expand?	*141*
Creation of matter	*142*
Cosmic microwave background radiation	143
Uniqueness of Yilmaz theory predictions	143

13. A believable picture of our universe — 144

The implications of the Yilmaz cosmology model	144
Matter derived from gravitational waves	*145*
The local expansion of the universe	*146*
The second law of thermodynamics	*146*
Contents of the universe	149
The "observable" Yilmaz universe	*149*
Is the universe infinite?	*150*
Religious and philosophical implications of our picture of the universe	151
The Biblical story of Creation	*151*
Our picture of the universe	*152*

APPENDICES

A. Density and matter in the universe — 154

Luminous density of the universe	154
Dark matter	154
The source of dark matter	155
Total mass in the universe	156
Mass density of the universe	157
Predicted density of matter	157
Theoretical mass of the universe	158
Rate of creation of matter	159
Size of the initial Big Bang universe	159

B. Cosmic microwave background radiation — 160

Glossary — 164

Bibliography — 168

Index — 171-176

Preface

"How were we created?" and "How were our world and our heavens created?" The book addresses these ageless questions.

It describes the creation of life on earth, since the earth cooled 4.3 billion years ago. After billions of years of microscopic life, seaweed appeared 1.8 billion years ago, and animals appeared 600 million years ago. The evolution of vertebrates is traced, from the first fish (510 million years ago) to the emergence of humanity.

We examine the creation of our sun and our earth. Gravity caused a cloud of hydrogen to collapse, heating the gas until nuclear fusion was ignited, 5 billion years ago. Our sun began to shine. A disk of gas and dust remained around the sun, and from this disk, the earth and other planets were created.

Our sun will continue to shine for another 6 billion years, until the nuclear fuel is exhausted. Then the sun will shrink to become a white dwarf star, glowing white hot from the energy released by gravity. When this white dwarf reaches to the size of the earth, it will stop shrinking, and will cool to become a dead and dark body called a black dwarf star.

Most stars have life cycles similar to our sun, but the power radiated and the lifetime vary greatly with the star's mass. However, stars with more than 10 times the sun's mass end their lives in a radically different manner. They die in an explosion called a supernova, which shines for a month with the brilliance of billions of suns. After the supernova, an extremely dense star remains, called a neutron star.

Our sun lies within our Milky Way galaxy, which is similar to the Whirlpool galaxy on the front cover. Our galaxy contains 100 billion stars, and is 100 thousand light years in diameter. As we look beyond our galaxy, we see billions of similar galaxies, each with many billions of stars. Then we observe the greatest mystery of all. Our universe is expanding. Galaxies are flying apart, as if emerging from an enormous explosion. What does this tell us about the creation of our universe?

Preface **xiii**

The Modern Big Bang Theory

For many years astronomers have claimed with absolute confidence that our universe was created in a "Big Bang" explosion about 15 billion years ago. They have concluded that the universe began as a ***singularity***, having nearly an infinite density of matter.

James Peebles, acclaimed by *Scientific American* as the "father of modern cosmology" predicted that the initial observable universe was ***"smaller than a dime"***. This constraint is generally accepted, but most Big Bang theories now support the 1981 ***"inflation"*** concept of Alan Guth, which claims that the initial size of the universe was ***microscopic.***

With the universe expansion, galaxies 15 billion light years away should recede at the speed of light and presumably cannot be seen. Hence our observable universe is considered to be a sphere 15 billion light-years in radius. Astronomical data show that the matter within this observable universe is equivalent to 20 billion times the mass of one trillion suns.

The concept that all of this matter in the observable universe could have been jammed into a volume "smaller than a dime", and possibly microscopic in size, drastically violates our common sense. Many question the validity of cosmology research, as illustrated by the following comment in the August 2001 *Scientific American* (p. 14):

> *"Whenever Scientific American runs an article on cosmology, we get letters complaining that cosmology isn't a science, just unconstrained speculation".*

The Gamow Big Bang Theory

A radically different version of the Big Bang theory was proposed in 1947 by George Gamow, who was a leading physicist in developing the atomic bomb. Matter within the atomic nucleus has a density of one billion tons per teaspoon, which is also the density of a neutron star. Gamow considered this to be the greatest possible density of matter, and he postulated that our universe began with this density at the Big Bang.

If we apply astronomical data to the Gamow postulate, our present observable universe at the instant of the Big Bang would have been about the size of the orbit of Mars around the sun. Although the Gamow Big Bang concept was once widely endorsed, in the 1960's the modern Big Bang theory replaced it with its unbelievable singularity concept.

The Einstein General Theory of Relativity

When computers became widely available in the 1960's, scientists began to use them to study the Einstein General theory of Relativity. These studies were applied to cosmology, which is about the only area that can use this expertise. With a computer, the extremely complicated Einstein equations could be solved in a manner unheard of in Einstein's day. This work stimulated a vast Big Bang research effort involving hundreds of scientists.

Computer studies of the Einstein theory predicted that the universe began as a singularity, and so the singularity became an essential feature of the modern Big Bang theory. Cosmologists are using the great prestige of Einstein to support their singularity claims, but are drastically violating the philosophy that Einstein demanded throughout his career.

The concept that the Einstein theory predicts a singularity was first proposed in a 1939 paper, which concluded that a star with sufficient density to form what was later called a *black hole* would contract "indefinitely". The star would shrink to become a singularity having an infinite density of matter. *Einstein absolutely rejected the black-hole singularity; he insisted that, "singularities do not exist in physical reality". After that rejection by Einstein, no scientist claimed that General Relativity predicted a singularity while Einstein was alive.*

In 1945, Einstein recognized that his theory seemed to imply a singularity at the birth of the universe. Einstein flatly rejected this interpretation, stating that his theory would not be valid under "very high density of field and of matter" and so cannot be used to predict a physical singularity.

The Yilmaz Theory of Gravity

The principles that Einstein established in developing General Relativity were sound. However, Einstein derived the formula that specifies his theory in an intuitive manner. That formula has a flaw, and the physically impossible singularity predictions of modern Big Bang cosmology are due to that flaw.

In the 1950's, Huseyin Yilmaz studied the Einstein theory as part of his PhD research at the Massachusetts Institute of Technology. He examined an approximate calculation that Einstein had made and discovered that he could implement it exactly. This yielded an exact solution to Einstein's Relativity principles and resulted in the Yilmaz theory of gravity. Since the Yilmaz theory applies the principles of General Relativity, it is a refinement of the Einstein theory.

The first paper on the Yilmaz theory was published in 1958 in the prestigious *Physical Review*, and after that Yilmaz has published numerous scientific papers to extend his theory. Yilmaz has demonstrated that his theory has profound mathematical integrity. The Yilmaz theory refutes the physically impossible singularity predictions derived from the Einstein theory. Since the Yilmaz theory is a refinement of the Einstein theory, it has proven that the basic principles of General Relativity are inconsistent with singularities.

Big Bang cosmologists reject the Yilmaz theory, which refutes their singularity predictions. The Yilmaz theory is easy to apply, and so would reduce to obsolescence the elaborate computer programs that have been developed to solve the Einstein equations.

The Steady-State Universe theory. In 1948, the famous physicist Fred Hoyle proposed this theory, which postulated that our universe is infinitely old and that diffuse matter is continually created to offset the universe expansion. The theory was strongly supported until *cosmic microwave radiation* was discovered in 1965, which had been predicted by Gamow. This microwave radiation was claimed to be the cooled relic of optical radiation emitted soon after the Big Bang. The Hoyle theory could not explain this radiation, and so was abandoned.

The Yilmaz cosmology model. The Yilmaz theory yields a cosmology model that is similar to the Steady State Universe theory, but does not have its limitations. As shown in Appendix B, the Yilmaz cosmology model predicts cosmic microwave radiation much more accurately than does the Big Bang theory.

The Yilmaz cosmology model makes the simple postulate that our universe has a constant average density of matter that does not change with time. Yilmaz found to his surprise that his model predicts an expanding universe. The expansion of the universe is a natural relativistic effect that is directly caused by gravity. How can gravity, which normally causes masses to attract, force the universe to expand? This remarkable prediction is explained in Chapter 12.

Philosophy of the Book. To study gravitational effects in the universe, we must understand Relativity theory. The book explains the Einstein and Yilmaz theories in a simple manner, which provides a coherent basis for exploring our universe. The insights that the reader gains from a journey through this book should be a rewarding adventure in the pursuit of the eternal Mystery of Creation.

Figure 1-1: The M51 Whirlpool galaxy, which is 35 million light years away, resembles our own Milky Way galaxy. Its smaller companion galaxy is NGC5195.

Chapter 1

Introduction

Since the dawn of human awareness, we have wondered, "How were we created?" and "How were our earth and heavens created?" The issues are so fundamental to our understanding of life that the Bible starts with its explanation: "In the Beginning, God created the heavens and the earth". Now let us see what the scientists say.

Out Mystical Milky Way

To gain insight into this mystery, look up at the night sky. Nearly every summer I visit a rural location that is far from city lights. I am always excited to view the sky on a clear moonless night, to see countless stars shining in wondrous majesty. The most inspiring feature to me is that pale white pathway across the sky, called the Milky Way. The Milky Way encircles the celestial sphere, dividing it into nearly equal hemispheres. It is sad that so many people today have never seen our mystical Milky Way, because it is obscured by sky glow due to the reflection of electric lights from the atmosphere.

The Milky Way is our view of 100 billion stars that form our Milky Way galaxy, which is 100 thousand light-years in diameter. A light-year is the distance that light travels in one year, which is 10 trillion kilometers. Our galaxy is similar to the Whirlpool galaxy in Fig. 1-1 and on the front cover. Our sun lies 2/3 of the distance from the center to the circumference. (The Whirlpool galaxy is 35 million light-years away, and is 2 degrees from the end of the Big Dipper handle.)

Prior to 1900 it was generally believed that our Milky Way galaxy was the whole universe. Then astronomers developed means of measuring stellar distances and discovered that many of the fuzzy astronomical objects called "nebulae" were actually distant galaxies containing billions of stars like our own Milky Way. Our Milky Way galaxy is merely one out of billions of galaxies that comprise our universe. The size of our universe seems beyond comprehension.

Yet there is something even more unbelievable about our universe. The universe seems to be flying apart. Galaxies are moving away from us at velocities approximately proportional to distance. What can this mean? What does it tell us about how our universe was created?

But before we consider the great mystery of our universe, let us begin with the simpler issues: the creation of our sun and our earth, and the creation of the stars.

The Creation of Our Sun and Our Earth

The studies of nuclear power have explained the process that generates the enormous energy radiated by our sun. Our sun was created 5 billion years ago when a cloud of hydrogen gas contracted because of gravity. The gravitational energy released by the contraction heated the gas until nuclear fusion was ignited, and our sun began to shine. "And God said, 'let there be light', and there was light".

Nuclear fusion generates the awesome power of the hydrogen bomb. Four hydrogen atoms are fused to form one helium atom. The mass is slightly reduced, and this loss of mass releases an enormous amount of energy.

A disk of gas and dust remained around the sun, and from this disk our earth and other planets were created. As the Bible says, "In the Beginning," . . . "The earth was without form and void, and darkness was upon the face of the deep".

The creation of our solar system is described in *Story* [4], Chapter 3, which shows that the formation of a solar system is a normal aspect of a star's development. Consequently many stars in out Milky Way galaxy probably have planets like earth that can support life.

Life Cycles of Our Sun and Other Stars.

Our sun has been generating energy for 5 billion years by fusing hydrogen to form helium, and will continue for another 5 billion years until the hydrogen is exhausted. Then it will fuse the helium to form carbon for another billion years. When the helium is exhausted, nuclear fusion will permanently end, and our sun will slowly collapse to form a white dwarf star, glowing white hot from the energy released by gravity. When the sun reaches the size of the earth, it will stop shrinking, and will gradually cool to become a dead body called a black dwarf star.

Stars with less than 8 times the mass of our sun have life cycles similar to our sun, but the rate at which they progress through their lives varies greatly with mass. A star with twice the mass of our sun radiates 16 times as much power, and its lifetime is 1/8 that of our sun. A star with half the mass of our sun radiates 1/16 times as much power, and its lifetime is 8 times that of our sun.

When a massive star with more than 10 times our sun's mass runs out of nuclear fuel, it ends its life in a dramatic **supernova** explosion. In this process each atom of the star suffers catastrophic collapse.

To understand this process, let us consider the structure of an atom. An atom is made up of 3 elementary particles: the electron, the proton, and the neutron. The electron has a negative electrical charge, and the proton has an equal and opposite positive electrical charge. The neutron has no electrical charge. The proton and the neutron have about the same mass, which is about 1840 times the mass of the electron. Protons and neutrons are contained within a tiny nucleus, around which orbits a cloud of electrons. The volume of the electron cloud is trillions of times greater than the volume of the tiny nucleus.

When an atom collapses in a massive star, the electrons are forced into the protons to form neutrons, and so the electron cloud is eliminated. Each atom shrinks to the size of its nucleus, and the star consists entirely of tightly packed neutrons. The resultant star is called a neutron star. The density of a neutron star is enormous. One teaspoon weighs one billion tons. Although this density may seem unbelievable, the same density exists here on earth within the nucleus of every atom.

This catastrophic gravitational collapse releases tremendous energy, which blows the outer portion of the star apart in a supernova explosion. The supernova shines with the brightness of more than one billion suns for about a month.

In the supernova explosion, neutrons and neutrinos are fired into the atoms of the outer portions of the star, producing nuclear reactions that create heavy elements. These heavy elements form dust particles that are scattered across the galaxy by the supernova explosion. The dust particles are gathered in the hydrogen clouds that form subsequent stars, and produce the solid matter from which planets like earth are created.

Chapter 2 explains in detail how our sun and other stars were created, and how they will eventually die. It also discusses the creation of life on earth.

4   *The Mystery of Creation*

The Universe that Lies beyond Our Milky Way

When we look outside our Milky Way galaxy to the vast universe that lies beyond, we see billions of galaxies that extend to the limits of our telescopes. Each of these galaxies contains many billions of stars.

Then we observe the astounding expansion of our universe. Galaxies are flying away from us at velocities approximately proportional to distance. The universe expansion was discovered by astronomer Edwin Hubble in 1929, and is called the Hubble expansion. Since Hubble's discovery, postulate after postulate have been proposed to explain this perplexing expansion of the universe.

The ratio of galaxy velocity divided by galaxy distance is called the Hubble constant. Hubble determined the velocity of a galaxy by measuring the redshift in the galaxy spectrum, which is the shift of the spectral lines toward red wavelengths

Based on recent measurements of the Hubble constant, this book assumes a universe expansion rate of 20 kilometers per second (km/sec) per million light-years of galaxy distance. For this expansion rate, a galaxy 15 billion light-years away should recede from us at the speed of light, 300,000 km/sec, and so should not be observable.

This implies that 15 billion light-years is our observational limit. Our observable universe is considered to be a sphere with a radius of 15 billion light-years; stars beyond this sphere presumably cannot be seen.

Meaning of the Hubble Expansion

The expansion of our universe leads us to the ultimate mystery: How was our universe created? How do we explain this enigmatic expansion of our universe?

The galaxies of our universe appear to be emerging from an enormous explosion. This is the obvious interpretation of the Hubble expansion, and is the answer that is believed today by nearly all astronomers. For many years astronomers have been claiming with absolute confidence that our universe began about 15 billion years ago in an explosion called the "Big Bang".

But there are other possible explanations of the Hubble expansion. The major alternatives are:

(a) The universe expansion is apparent. Hubble measured galaxy velocity from a redshift in the galaxy spectrum, but effects other than velocity can also produce redshift. Some scientists believe that the Hubble redshift is not due to velocity, and so our universe is not actually expanding.

(b) The Steady-State Universe theory. In 1948, the noted astrophysicist Fred Hoyle proposed this cosmological theory, which postulates that matter is created to compensate for the Hubble expansion, so that the average density of matter remains constant. The required creation of matter is far too small to be directly observed. The created matter would form new galaxies and stars, so that the universe continually changes in detail. Our universe is infinitely old, but remains eternally young from the creation of new stars and galaxies.

The Steady-State Universe theory ran into difficulties in the late 1960's, and was gradually abandoned. However, a new cosmology theory is described in this book, which incorporates the major principles of the Steady-State Universe theory, but does not have the limitations of the original theory. This new theory will be discussed later.

There are actually two radically different versions of the Big Bang theory, which are: (1) the Big Bang theory publicized by George Gamow in the 1940's and 1950's; and (2) the modern Big Bang theory based on the singularity principle, which evolved in the 1960's. Let us examine these two Big Bang theories.

The Gamow Big Bang theory. George Gamow was a renowned nuclear physicist who was a leader in the development of the atomic nuclear bomb. After World War II, Gamow wrote popular books on physics and astronomy that were widely read. In one of these [16], he postulated that our universe began billions of years ago as a highly dense mass, which had the density of the atomic nucleus (one billion tons per teaspoon). Gamow considered this to be the greatest possible density of matter. It is also the density of a neutron star, which consists of tightly packed neutrons. Gamow postulated that this extremely dense mass exploded with a "Big Bang" billions of years ago,.

The modern Big Bang theory. The General theory of Relativity presented by Albert Einstein in 1916 has extremely complicated equations that can be applied analytically only to very simple physical models. In the mid 1960's, powerful computers became widely available, and hundreds of scientists began to use computers to study the Einstein theory. With computers, the Einstein equations could be solved in a manner unheard of in Einstein's day. Because of the great prestige of Einstein, this computer research was professionally very rewarding.

About the only area in which these computer studies of the Einstein equations could be applied was cosmology. Consequently the computer studies resulted in an enormous amount of cosmology research, which invariably led to the Big Bang theory. Cosmological computer studies of the Einstein theory predicted that the universe began as a "singularity", which ideally means an infinite density of matter.

This computer research based on the Einstein equations resulted in the modern Big Bang theory, which claims that the universe began as a "singularity" having almost zero size and nearly infinite density. The density that was predicted vastly exceeds the density of nuclear matter that was postulated by the Gamow Big Bang theory.

Thus, there are four radically different concepts for explaining the Hubble expansion of the universe, which are:

(1) *The Gamow Big Bang theory*, postulating an initial universe with the density of nuclear matter;
(2) *The modern Big Bang theory*, postulating an initial "singularity" universe having nearly infinite density;
(3) *The apparent expansion theory*, postulating that the universe is not expanding, and that the Hubble redshift is not due to velocity;
(4) *The Steady-State Universe* concept, postulating an infinitely old universe with matter created to offset the Hubble expansion.

Initial Size of the Big Bang Universe

Let us examine the initial size of the observable universe predicted by the two versions of the Big Bang theory. As shown in Appendix A, astronomical data indicates that our observable universe (30 billion light-years in diameter) contains matter equivalent to about 20 billion times the mass of one trillion suns. Let us imagine that the observable universe is squeezed to form a sphere with the density of water. Appendix A shows that this sphere would be 4.4 light-years in diameter, which is about the distance (4.2 light-years) to the nearest star.

Let us imagine that this water-density sphere is squeezed further to achieve the density of nuclear matter (one billion tons per teaspoon). The sphere would then have a diameter of 700 million km, which is 1.5 times the diameter of the orbit of the planet Mars around the sun.

A neutron star has the density that exists within an atomic nucleus, and consists of tightly packed neutrons. Gamow considered this to represent the greatest possible density of matter. There is extensive physical evidence supporting this conclusion.

In summary, the Gamow Big Bang theory postulated that our universe began with the density of nuclear matter. According to this postulate, our observable universe would have begun as a sphere having 1.5 times the diameter of the orbit of Mars around the sun.

An extensive discussion of the modern Big Bang theory is given in the January 2001 *Scientific American*, which called astrophysicist James Peebles the "father of modern cosmology" (page 37). The October 1994 *Scientific American* (page 53) presented an article by Peebles and others, which began with:

> "At a particular instant, roughly 15 billion years ago, all of the matter and energy we can observe, **concentrated in a region smaller than a dime**, began to expand and cool at an incredibly rapid rate."

Nearly all modern cosmologists endorse this claim by Peebles that the observable universe at the Big Bang was "smaller than a dime", and many believe it was very much smaller.

Most of the recent Big Bang theories support the *"inflation"* concept proposed by Alan Guth in 1981, which postulates that the initial universe was microscopic in size. Dickinson [24] (page 118) gives a pictorial representation of the Big Bang singularity that includes Guth's inflation postulate. **Dickinson states that during the inflation period, the observable universe "expands from one trillionth the size of a proton to the size of a baseball".** Since Dickinson's book is a third edition of a classic astronomy book, with excellent illustrations and explanations, its statements obviously portray the general thinking of astronomers.

The Peebles claim that the initial observable universe was "smaller than a dime" requires a density of matter that exceeds that of the Gamow postulate by a factor of 10^{40}. This factor means one followed by 40 zeros (10 billion, times 10 billion, times 10 billion, times 10 billion)! The Guth postulate requires a density of matter that exceeds the density claimed by Peebles by a factor of 10^{76} (one followed by 76 zeros)!

These singularity claims of modern Big Bang cosmologists are so extreme they are devoid of physical reality. There is extensive physical evidence showing that the density of nuclear matter postulated by Gamow represents the greatest possible density of matter.

Einstein's Rejection of the Singularity

Modern cosmologists are using the Einstein General theory of Relativity to support their unbelievable singularity predictions. Nevertheless, these claims radically violate the philosophy that Einstein demanded throughout his career.

Einstein recognized, as did Gamow, that nuclear matter has the greatest possible density of matter. The concept that our universe could be compressed to a density that is many, many times greater than the density of nuclear matter strongly conflicts with physical evidence. Einstein never accepted this physically impossible singularity concept.

In 1945, Einstein recognized that his theory seemed to imply a singularity at the birth of the universe. He flatly rejected this interpretation of his theory with the following [5] (page 129):

"Theoretical doubts [concerning the creation of the universe] are based on the fact that [at the] beginning of the expansion, the metric becomes singular and the density becomes infinite. . . In reality, space will probably be of a uniform character, and the present [relativity] theory will be valid only as a limiting case. . . **One may not therefore assume the validity of the equations for very high density of field and of matter, and one may not conclude that the 'beginning of the expansion' must mean a singularity in the mathematical sense.** *All we have to realize is that the equations may not be continued over such regions."*

In this quotation, Einstein stated that his theory could not be used to justify a physical singularity, because his equations would not apply under conditions of extreme density of matter. The thinking of Albert Einstein on such issues is discussed in a recent biography of Einstein, originally written in German by Folsing [23], which states (p. 381):

"Some of Einstein's admirers were tempted to see the general theory of relativity as a triumph of speculation over empiricism. This kind of misunderstanding made Einstein 'downright angry' [who said] 'This development teaches us something entirely different, indeed almost the opposite, namely that a theory, in order to merit confidence, must be based on generalizeable facts'. . . . To Einstein, facts were not only the starting point of his theory but also the keynote of any test of it."

1. Introduction

Einstein had extensive experience with physical experiments and was absolutely committed to the principle that theory must agree with physical evidence. The idea that the Einstein theory could be used to make predictions that grossly conflict with experimental evidence is a drastic violation of Einstein's scientific philosophy.

The first claim that the Einstein theory predicted a singularity was in a 1939 scientific paper, which concluded that an extremely dense star should collapse "indefinitely" until it shrinks to form a singularity. It was predicted that light cannot escape from such a star, and the star was later called a "black hole". Einstein replied with a paper that flatly rejected the black hole singularity. We will discuss the black hole later.

Einstein never accepted the black hole singularity or the Big Bang singularity. After Einstein's 1939 rejection of the black hole singularity, no scientist claimed that the Einstein theory predicted a singularity while Einstein was alive.

Limitations of the Einstein Theory

The fundamental cause of the confusion in cosmology is that modern cosmologists are treating the Einstein General theory of Relativity as absolute truth. However, as the above 1945 quotation shows, even Einstein recognized that his theory was limited, and could not be used to predict a physical singularity.

The recognition of limitations in the Einstein theory should not diminish our great respect for Einstein's genius. General Relativity has a broad foundation of sound scientific principles. However, the *Einstein gravitational field equation*, which specifies General Relativity, was developed by Einstein in an intuitive manner, after many years of searching for a general relativistic solution. The gravitational field equation is a tensor formula representing ten independent equations. Since experimental tests seemed to verify this Einstein formula, it was widely accepted as being valid.

The Einstein formula was so complicated, it could only be applied to very simple cases during Einstein's lifetime, and so its weaknesses were not apparent. When computers were applied to this formula, a decade after Einstein's death, they yielded physically impossible singularity predictions, which Einstein certainly would have opposed. Nevertheless, these singularity predictions have been solidly accepted by the scientific community, even though they strongly conflict with experimental evidence and with statements made by Einstein.

The Yilmaz Theory of Gravity

An answer to this confusion is provided by the theory of gravity that was published in 1958 by Huseyin Yilmaz in the *Physical Review* [Y1]. In the early 1950's, Yilmaz was studying General Relativity as part of his PhD research at the Massachusetts Institute of Technology. He examined an approximate calculation that Einstein made in developing General Relativity and discovered that he could implement it exactly. This yielded an exact solution to the principles of Relativity.

From this exact solution Yilmaz derived in a rigorous manner a different *gravitational field equation*, which is the foundation for the Yilmaz theory of gravity. Since the Yilmaz theory has applied the principles of General Relativity, it is a refinement of the Einstein theory. The Yilmaz theory does not allow a singularity and so has proven that the basic principles of the Einstein theory are inconsistent with singularities.

Yilmaz sent his analysis to Einstein, but Einstein died before he could read it. In 1958, Yilmaz published the first paper on his theory in the prestigious *Physical Review*, and since then has published numerous scientific papers to extend his theory.

Because the Yilmaz theory has an exact solution, it is very much easier to use than the Einstein theory. The elaborate computer analyses that have been developed to apply the Einstein theory are not needed with the Yilmaz theory. If the Yilmaz theory should become widely accepted, these computer techniques would become obsolete. Besides, the Yilmaz theory has refuted the physically impossible singularities that have been derived from the Einstein theory. Therefore, it is not surprising that the Yilmaz theory is opposed by the army of Big Bang cosmologists that are engaged in computer studies of the Einstein theory.

The Yilmaz Cosmology Model

In the first paper on his gravitational theory, Yilmaz applied his theory to a simple cosmology model, which assumed that the universe has a constant average density of matter that extends to infinity and does not change with time. To his surprise, Yilmaz discovered that his model predicts an expanding universe. Relativistic gravitational effects should make the universe expand, just as Hubble observed.

How can gravity, which always causes masses to attract one another, force the universe to expand? A simple physical explanation of this prediction is given in Chapter 12.

Since the first paper on his theory, Yilmaz has ignored cosmological applications of his theory, because he found cosmology to be speculative. The author has extended the cosmology model presented by Yilmaz, and has found that it yields some amazing predictions.

The Yilmaz cosmology model predicts a universe that is similar in many respects to the Steady State Universe theory that was presented by Fred Hoyle in 1948. However, the Yilmaz cosmology model does not have the limitations that led to the abandonment of the original Steady-State Universe theory.

The Yilmaz cosmology model portrays a universe that is infinitely old. Energy radiated from stars is converted into diffuse matter throughout space that compensates for the Hubble expansion. This diffuse matter forms new galaxies and stars, so that the universe continually changes in detail. Nevertheless, the overall form of the universe stays the same. The universe has always existed, looking more or less as we see it today, and will always continue to exist.

The energy radiated from stars consists of electromagnetic radiation (light, X-rays, radio waves, etc.) and gravitational waves. Gravitational waves have been predicted by both the Einstein and Yilmaz theories. It is postulated that the matter within black dwarf stars and neutron stars is continually radiated away in the form of gravitational waves.

The Yilmaz cosmology model predicts an exciting new picture of the universe, which does not violate any physical laws. Since it is based on the Yilmaz theory of gravity, which has a profound mathematical foundation, it deserves serious consideration. Let us examine its predictions

The Universe Predicted by the Yilmaz Theory

Many different cosmology theories have been **postulated** since the enigmatic Hubble expansion of the universe was discovered in 1929. The following discussion deals with cosmological **predictions**, not **postulates**. A *postulate* is a guess, which may or may not be correct. A *prediction* is derived from a more basic theory, and so its justification is as strong as the validity of the basic theory.

The Yilmaz gravitational theory has a rigorous and profound mathematical foundation, which incorporates the General Relativity principles established by Einstein. The Yilmaz theory is tightly constrained and so can yield only a single solution for a given physical model. (In contrast, multiple solutions for the same physical model can be derived from the Einstein equations.)

The Yilmaz cosmology model makes the simple postulate that the universe has a constant average density of matter that extends to infinity and does not change with time. Since the solutions of the Yilmaz theory are unique, the predictions derived from this simple physical model have firm scientific significance.

Let us examine the ***predictions*** that have been derived from the Yilmaz cosmology model. These predictions were calculated in a rigorous manner from the equations of the Yilmaz gravitational theory.

The Yilmaz cosmology model predicts that the universe must expand. The Hubble expansion is a relativistic effect that is directly caused by gravity. The universe expansion rate depends on the average density of matter of the universe.

For the universe to expand at our assumed Hubble rate (20 km/sec per million light-years), the average density of matter in the universe must be equivalent to ***4.8 hydrogen atoms*** per cubic meter. This value is calculated in Appendix A. That appendix also shows that the measured average density of matter of the universe derived from astronomical data is equivalent to ***2 hydrogen atoms*** per cubic meter. Considering the large sources of error involved in these calculations, this represents very good agreement between measured and theoretical values of the average density of matter.

Although our model assumes that the ***true distance*** of a galaxy can theoretically extend to infinity, the model predicts that the maximum ***apparent distance*** of a galaxy is finite. Because of the relativistic distortion of space produced by gravity, the apparent distance of any galaxy cannot exceed 18.8 billion light-years.

As the true distance of a galaxy increases, its apparent velocity gets closer and closer to the apparent speed of light, but never quite reaches it. Because of this effect, there is strong redshift of the light from far distant galaxies. The galaxy light is shifted to very low frequencies, to produce strong radiation at microwave frequencies. As shown in Appendix B, this effect accurately explains the cosmic microwave background radiation.

When cosmic microwave background radiation was discovered in 1965, Big Bang proponents claimed that this microwave radiation is the "smoking gun" that proves the validity of the Big Bang theory. The Yilmaz theory solidly refutes that claim.

The apparent velocity of a galaxy gets closer and closer to the apparent speed of light, as the true distance of a galaxy increases. Nevertheless, beyond 12.5 billion light years of true galaxy distance, the true velocity of a galaxy decreases with distance, because the apparent speed of light decreases.

The true velocity of a galaxy becomes very small for galaxies at large true distances. This indicates that over very large distances the universe does not expand. The over-all size of the universe does not increase with time. *Even though the universe expands locally about every point in the universe, the universe does not get any bigger.*

This relativistic property caused by gravity may seem strange, but so are all relativistic effects. As Chapter 6 will show, the Special Relativity effects due to velocity that were predicted by Einstein in 1905 seemed unbelievable when they were first presented. For example, Einstein showed that an object does not have an absolute length; the length of an object depends on the velocity of the observer that is measuring it. Although we have become accustomed to these relativistic effects, they are still strange. *The concept that our universe does not get any bigger as it expands everywhere should be no less acceptable than these earlier relativistic predictions.*

When the Hubble expansion of the universe was discovered in 1929, it was regarded as a deep mystery that must be answered. Since then countless postulates have been proposed to explain this perplexing enigma. The Yilmaz gravitational theory tells us that the Hubble expansion is not an enigma; it is an essential requirement if an infinitely old universe is to remain eternally young. *The Hubble expansion of the universe allows new matter to be created, which forms new stars and galaxies. Matter from old dead stars is continually recycled by the universe expansion to create new stars.*

Here is a fresh picture of the universe. It is not a product of speculation. It was calculated rigorously from the equations of the Yilmaz theory, which has a profound scientific foundation and is a refinement of the Einstein theory. Is this prediction of the Yilmaz theory correct? As a minimum, it provides an exciting new concept for explaining the creation of our universe.

Understanding the Einstein and Yilmaz Theories

It should be clear from the preceding discussion that to explore the mysteries of cosmology one must understand the Einstein and Yilmaz theories. Even though the Einstein equations are very complicated, this book shows that the principles of the Einstein and Yilmaz theories can be explained in a simple manner.

Chapter 2

Creation of the Sun, Earth, and Stars

This chapter describes the creation of our sun and earth and the creation of life on earth. It then examines the life cycle of our sun and other stars, showing how these stars will eventually die. More extensive discussions of these issues are given in *Story* [4] (Chapters 2 and 3).

The Creation of Our Sun and Our Earth

Let us start our story by examining the creation of our sun and our earth. Our sun was created about 5 billion years ago when a cloud of hydrogen was drawn together by the force of gravity. As the cloud collapsed, gravitational energy was released, which heated the cloud until the center reached a temperature of 15 million degrees Celsius. The temperature and pressure were then sufficient to ignite nuclear fusion, which is the process behind the awesome power of the hydrogen bomb.

In this nuclear fusion process, four hydrogen atoms are fused together to form one helium atom. This reduces the mass by 0.7 percent, and the mass that is lost is converted into an enormous amount of energy. For every gram of hydrogen converted into helium, 180,000 kilowatt-hours of energy are released. Our sun converts 650 million tons (650 trillion grams) of hydrogen into helium every second. Our sun has been converting hydrogen into helium for five billion years and will continue to do so for another five billion years, until the hydrogen fuel is exhausted.

A disk of gas and dust surrounded the sun, and from this disk came the earth and the other planets. The dust was produced by stars that exploded earlier as supernovas. Material in the disk congregated to form our earth as a molten body, 4.6 billion years ago. Since the oldest earth rocks are 4.3 billion years old, the crust of the earth probably solidified 4.3 billion years ago.

Our trips to the moon have provided information concerning the early life of the earth, because the earth and moon have received similar meteorite bombardment. Moon rocks from craters on the moon have shown that until 3.8 billion years ago the moon and the earth were heavily bombarded by meteorites. Therefore until 3.8 billion years ago our earth had a hellish environment, so severe that very little of the life that we know today could have existed.

When the earth was a molten body, most of the water may have boiled away. However, meteorites contain a large amount of water, and so the water that fills our oceans probably came primarily from meteorites. Life on earth began in our oceans.

Early Microscopic Life

The first definite signs of life appeared 3.6 billion years ago in the form of *stromotolites*, which are fossils of colonies of cyanobacteria. These bacteria perform photosynthesis, which absorbs sunlight to create food. Photosynthesis is the process implemented in plants, which combines the energy from sunlight, the hydrogen from water, and the carbon from carbon dioxide to synthesize carbohydrates. Photosynthesis is performed by a chemical called chlorophyll.

Scientists have only recently found the key to this mystery of how life began on earth, with the discovery of volcanic vents on the ocean floor. These vents are surrounded by extensive life feeding on microscopic cells that derive nourishment from hydrogen sulfide and other chemicals emanating from the volcanic vents. These microscopic cells were first thought to be bacteria, but DNA studies have shown them to be radically different. They have been given the name *archaea*. Archaea are similar to bacteria but can tolerate very harsh environments, including very high temperatures. Archaea were first discovered in the late 1970's in hot springs, such as those in Yellowstone National Park.

The primary building blocks of life are amino acids. The volcanic vents on the ocean floor have the energy and chemicals that could have synthesized the basic amino acids by inorganic means. Consequently, many biologists believe that archaea were the first organisms on earth. Archaea could have flourished in the hostile environment, 4.3 to 3.8 billion years ago, when the earth was heavily bombarded by meteorites.

The membranes of archaea are chemically different from those of bacteria. Chemical evidence has identified archaea in rocks 3.8 billion years old. However, archaea may have evolved soon after the earth's crust solidified 4.3 billion years ago. Microscopic archaea cells probably were the only life on earth until the meteorite bombardment ended 3.8 billion years ago. Then the more delicate bacteria could survive.

The first bacteria probably fed on archaea, which derived their energy from chemicals in volcanic vents. In time, certain bacteria began to synthesize chlorophyll and thereby implemented photosynthesis. Sunlight was now used as the basis for creating food that supports life. Today the energy derived from sunlight dominates life processes on earth, so much so that biologists have only recently discovered that the energy that is feeding archaea comes from an entirely different source.

Archaea and bacteria are tiny cells that lack nuclei. For about 1.5 billion years these simple cells, first archaea and then bacteria, comprised the life on earth. Then 2.7 billion years ago a new type of cell evolved, called the eukaryote. A eukaryote cell is much larger than an archaea or bacteria cell and contains a nucleus. More complicated biological processes occur within the eukaryote cell, and it allowed the evolution of multi-celled organisms. All multi-cell organisms consist of eukaryote cells. Strange as it may seem, the DNA of eukaryotes is closer to that of archaea than to bacteria.

The Evolution of Multi-Celled Plants and Animals

Table 2-1 lists the primary events that have been involved in the development of life on earth. We have just discussed events (1) to (6). For billions of years, life on earth consisted of single-celled organisms. Some of the eukaryote cells performed photosynthesis, and were called algae. Multi-celled organisms began when algae calls formed the first marine plant (seaweed), which appeared 1.8 billion years ago (7). Terrestrial plants evolved much later.

Photosynthesis performed by cyanobacteria, single-celled algae and seaweed, released oxygen into the atmosphere as a byproduct. The oxygen finally reached sufficient concentration to support animal life. Jellyfish and other soft-body animals appeared about 600 million years ago (8). An explosion of animal life appeared in the oceans 545 million years, at the start of the Cambrian geological period (9). Within 10 million years representatives from nearly all of the animal types (phyla) appeared. One of these was a primitive representative of our own phylum, the Chordates, which includes the vertebrates (animals with backbones).

The first vertebrate (10) was a jawless fish, which appeared 510 million years ago. Fishes with jaws (11) evolved 70 million years later. Sharks (13) appeared 415 million years ago, and modern bony fish appeared 400 million years ago (14).

Table 2-1: *Major events in the evolution of life on earth*

Event	Years Ago
(1) earth formed	4.6 billion
(2) earth crust hardened	4.3 billion
(3) meteorite bombardment ceased	3.8 billion
(4) evidence of archaea	3.8 billion
(5) cyanobacteria colonies	3.6 billion
(6) eukaryotes	2.7 billion
(7) seaweed	1.8 billion
(8) first animals	600 million
(9) explosion of animal life (Cambrian)	545 million
(10) first vertebrate (jawless fish)	510 million
(11) jawed fish	440 million
(12) first terrestrial plants (mosses)	430 million
(13) sharks	415 million
(14) modern bony fish	400 million
(15) ferns and related plants	400 million
(16) amphibians	360 million
(17) reptiles	335 million
(18) conifer and cycad plants	300 million
(19) dinosaurs	225 million
(20) mammals	215 million
(21) Age of Dinosaurs	205-65 million
(22) flowering plants	140 million
(23) birds	135 million
(24) Age of Mammals	65 million to present

The first terrestrial plants (12) appeared 430 million years ago. They were low plants similar to mosses. Ferns and related plants (15) appeared 400 million years ago. The fern-like plants had vascular systems for conducting fluids, which allowed them to grow tall, some reaching 150 feet. Fern-like plants formed massive coal deposits that we now use for fuel. All of these plants reproduced by spores.

Conifer and cycad plants (18) appeared 300 million years ago. These plants have seeds. Flowering plants (22) evolved about 140 million years ago, in the middle of the Dinosaur Age.

Amphibians (16) moved onto the land 360 million years ago.

18   The Mystery of Creation

Amphibians (which include frogs) lay small fragile eggs in the water, like fish eggs. They begin their lives like fish as tadpoles, developing lungs and feet as they mature.

The first reptile appeared 335 million years ago (17). Reptiles have large amniotic eggs, like bird eggs, which can hatch on land.

Dinosaurs (19) and mammals (20) appeared about the same time. For 140 million years, from 205 to 65 million years ago, the dinosaurs dominated the land during the Age of Dinosaurs (21). In this period, few mammals were larger than a mouse. Birds (23) appeared in the middle of the dinosaur period, and apparently evolved from dinosaurs. Birds competed with pterodactyls, which were flying reptiles closely related to dinosaurs. Pterodactyls existed throughout the dinosaur period, and died with the dinosaurs.

Sixty-five million years ago, an asteroid about 10 kilometers in diameter hit the earth near the Yucatan peninsula in Mexico, and caused a catastrophe that killed off the dinosaurs. Mammals and birds survived, probably because many were small and could eat a variety of foods, such as seeds, roots, and mushrooms. Without competition from the dinosaurs and pterodactyls, mammals and birds evolved rapidly, and our modern Age of Mammals (24) began.

The Emergence of Humanity

Table 2-2 lists the major events associated with the emergence of humanity. Humans belong in the *primate* order of Mammals. The first primates (1) appeared 55 million years ago. They were *prosimians*, which include lemurs. The first monkeys (2) appeared 35 million years ago, and the first apes (3) appeared 25 million years ago.

About 6 million years ago, Australopithecus (4) appeared in Africa. It was similar to a chimpanzee but could walk upright. Climate changes had thinned our jungle trees, and Australopithecus needed to travel over appreciable distances between trees.

The first member of our own genus was Homo Habilis ("handy man") (5), who appeared 2.5 million years ago. He had an appreciably larger brain than Australopithecus, and made crude stone tools. Although chimpanzees and Australopithecus were primarily vegetarians, Homo Habilis ate meat regularly, and the stone tools were used to butcher the meat.

Homo Erectus (6) appeared 1.8 million years ago. He had a larger brain than Homo Habilis, and made more complicated stone tools. Homo erectus was about 6 feet tall, and had a skeleton similar to modern man, except for the head. The Homo Erectus skeleton had the flexibility to allow fast running. Australopithecus had a pot belly like a chimpanzee, and could walk but could not run. Homo Erectus migrated from Africa into Europe and Asia.

Table 2-2: Major events in the development of humanity

Event	Years Ago
(1) prosimians (first primates)	55 million
(2) monkeys	35 million
(3) apes	25 million
(4) Australopithecus (upright ape)	6 million
(5) Homo Habilis	2.5 million
(6) Homo Erectus	1.8 million
(7) Archaic Homo Sapiens	600 thousand
(8) Neanderthal man	200 thousand
(9) anatomically modern humans	100 thousand
(10) modern human behavior	40 thousand
(11) Neanderthal extinction	35 thousand
(12) agriculture development	12 thousand
(13) first towns	10 thousand

The brain of Homo Erectus increased with time, and by 600 thousand years ago Homo Erectus had evolved into Archaic Homo Sapiens (7). Neanderthal Man (8) appeared about 200 thousand years ago. His brain was as large as modern man, but he had a sloping forehead. He was much stronger than modern man. The first anatomically modern man (9) appeared 100 thousand years ago.

For 60 thousand years, modern humans made the same stone tools as Neanderthal man, and they apparently lived similar lives. However, Neanderthal man probably lived in the ice-age regions of Europe, while modern humans lived in warmer regions.

Then, suddenly, about 40 thousand years ago, modern humans started making much more sophisticated implements in stone and bone, which included many with an artistic purpose, including beads, pendants, etc. This change represented a revolution in human behavior. (10) Soon thereafter, 35 thousand years ago, Neanderthal man became extinct (11).

Agriculture was invented 12 thousand years ago. (12) The earlier hunter-gather way of life could only support a sparse population. Agriculture allowed the development of towns and then cities. The first towns of significant size appeared 10 thousand years ago. (13). After that time, complex civilizations rapidly emerged.

Was There a Divine Spark in the Development of Humanity?

The preceding explanation of the evolution of life on earth, from early microbes to the development of modern humans, has applied the principles of natural selection. The natural selection process capitalized on random genetic mutations, augmenting those mutations that yielded advantageous characteristics. By this means the complex sophisticated life of today could have evolved from very simple life in a countless series of small changes that occurred over billions of years.

Is that all that there was? Was any divine purpose involved in this process? DNA analysis has demonstrated extreme similarity in the genetic codes of different species. It seems impossible to believe that the cells in the bodies of humans are not biologically related to the cells of other species.

However, this does not prove that there was no divine spark in the development of humanity.

The revolution of modern human behavior that occurred about 40 thousand years ago is a phenomenon that cannot be attributed to biological evolution. Modern human behavior was created far too suddenly. Since modern human behavior developed much too quickly to be explained by biological evolution, one can legitimately ask the question, "Was this development the result of Divine intervention?"

The Life Cycles of the Stars

Our sun has been generating energy by fusing hydrogen to form helium for five billion years, and will continue this process for another five billion years until the hydrogen fuel is exhausted. Then the sun will contract to achieve a temperature of 100 million degrees Celsius. A second nuclear fusion process will occur in which 3 helium atoms are fused to form one carbon atom. This process will reduce the mass by

0.07 percent, and will release 18,000 kilowatt-hours of energy per gram of helium. When the helium fuel is exhausted, nuclear fusion will permanently stop.

The white dwarf star. Our sun will become a white dwarf star, which gradually shrinks to the size of the earth, glowing white-hot from the energy released by gravity. When our sun reaches the size of the earth, it will stop shrinking and so will stop radiating energy. It will gradually fade to become a black dwarf star, the dead ember of a once brilliant star. It will have a density of about one ton per cubic centimeter, which is one million times the density of water.

One would not want to land a space ship on this black dwarf star. The force of gravity would be one million times that on earth, and the space ship with its inhabitants would be crushed.

There is abundant evidence that white dwarf stars (with their extremely high densities) actually exist. The brightest star in the sky is Sirius, which has a companion star that is a white dwarf. This white dwarf has been studied extensively. Its mass can be determined from its gravitational effect on the motion of Sirius. Sirius and its white-dwarf companion are 8.6 light-years away.

The basic life cycle that we have discussed for our sun is followed by all stars with less than 8 times our sun's mass, but the rate at which a star proceeds through this life cycle varies greatly with its mass. A star with twice our sun's mass generates 16 times as much power, and has 1/8 of the lifetime of our sun. A star with one-half of the sun's mass generates 1/16 as much power, and has 8 times the lifetime of our sun. Stellar mass varies from 1/12 of the sun's mass to 100 times its mass.

The supernova. A star with more than 10 times our sun's mass ends its life in a dramatic explosion called a supernova. A massive star begins by fusing hydrogen to form helium, and then fuses the helium to form carbon. However, with its greater mass, the star can achieve the temperature to generate the following elements by nuclear fusion: oxygen, neon, silicon, nickel, cobalt, and iron. Then the fusion stops, because the generation of elements heavier than iron does not release energy; it absorbs energy. The fusion processes from carbon to iron reduce the stellar mass by another 0.12 percent, and release 30,000 kilowatt-hours of energy per gram of carbon.

When this massive star runs out of nuclear fuel, catastrophic gravitational collapse suddenly occurs. The electrons of each atom are forced into the protons, converting them into neutrons. The electron cloud is eliminated, and each atom shrinks to the size of its nucleus. A **neutron star** is formed which consists entirely of tightly packed neutrons. The star has reached the maximum possible density of matter, 200 million tons per cubic centimeter. **One teaspoon weighs one billion tons.** Although this density may seem unbelievable, this same density exists here on earth within the nucleus of every atom.

When catastrophic gravitational collapse occurs, enormous energy is released, thereby causing the outer portions of the star to explode in a brilliant supernova, which shines with the brightness of about one billion suns for a month. In this explosion, neutrons and neutrinos are fired into the outer portions of the star, creating nuclear reactions that produce elements that are heavier than iron. These heavy elements, along with other elements created earlier in the star, are scattered as dust particles throughout the galaxy by the supernova explosion. These dust particles are picked up in the hydrogen clouds that form new stars, and produce the material from which solid planets like our earth are created.

The neutron star and the pulsar. The neutron star was originally only a theoretical prediction until the pulsar was discovered in 1968. A pulsar emits a radio signal consisting of pulses at precisely timed intervals. When first observed, some people thought that these pulses might be radio signals from intelligent beings on a distant planet. However, studies soon indicated that they are probably generated by rapidly spinning neutron stars.

A pulsar has been found at the location of a supernova that exploded in 1054 AD, and was recorded by Chinese astronomers. This pulsar is surrounded by the Crab Nebula, which is a cloud of gas produced by the supernova. This pulsar generates precisely timed radio pulses at 30 times per second.

Theory indicates that a pulsar is a rapidly spinning neutron star. The magnetic field of the neutron star radiates radio beams from north and south magnetic poles, directed along the magnetic axis. The energy creating this radio beam is derived from atoms being sucked onto the neutron star. Like the earth, the magnetic axis does not coincide with the spin axis, and so the radio beams spin in space as the neutron star rotates. When a radio beam passes the earth, we receive a radio pulse.

The pulsar star spins at the pulse frequency. A pulsar with a frequency of 30 pulses per second is a star rotating at 30 revolutions per second. For a star to spin at the frequency of a pulsar, it must be extremely compact. The only explanation that satisfies the pulsar characteristics is a neutron star. Therefore, even though a neutron star has very strange properties, there is strong evidence that neutron stars actually exist.

2. Creation of the Sun, Earth, and Stars

The Structure of the Atom

The processes involved in the life and death of stars become clearer when we examine the structure of the atom. All matter consists of combinations of about 100 different kinds of atoms. An atom is made of three elementary particles: the electron, the proton, and the neutron. The electron has a negative electrical charge. The proton has 1840 times the mass of the electron, and has a positive electrical charge, equal and opposite to that of the electron. The neutron has no electrical charge and has approximately the mass of the proton. All three elementary particles have been observed as separate particles outside the atom.

The protons and neutrons of an atom are contained within a very compact nucleus. The electrons orbit around the nucleus forming a spherical electron cloud that typically has about 50,000 times the diameter of the nucleus.

Densities of White Dwarf and Neutron Stars

Figure 2-1 illustrates the effect on the atom as a star is compressed to the density of a white dwarf star, and then to a neutron star. Before running out of nuclear fuel, our sun will consist primarily of carbon atoms. Diagram (a) shows the approximate dimensions of a carbon atom at normal pressure. A carbon atom has 6 protons (P) and 6 neutrons (N). The diameter of the carbon atom is about 0.25 nanometer, where a nanometer is one billionth of a meter. (The wavelength of visible light is about 600 nanometers.) The diameter D of the nucleus of the carbon atom is about 5 millionths of a nanometer.

Hence the diameter of the carbon atom (0.25 nanometer) is 50,000 times greater than the diameter of its nucleus, which is 5 millionths of a nanometer. The volume of the carbon atom exceeds that of the nucleus by the cube of 50,000, which is 125 trillion (125 million, million). As shown, the density of carbon is about 2 grams per cubic centimeter.

Heat generated by nuclear fusion produces pressure to offset the compression of gravity. When the nuclear fuel is exhausted, the pressure from heat will stop, and the sun will shrink to a white dwarf star. Its diameter will decrease by 100, and so its density will increase by the cube of 100, which is one million. Diagram (b) shows that an atom diameter is reduced from 0.25 nanometer to 0.0025 nanometer. Density is increased by one million, from 2 grams per cubic centimeter in (a) to 2 metric tons per cubic centimeter in (b). A metric ton is 1000 kilograms or one million grams, and is 10 percent greater than an English ton.

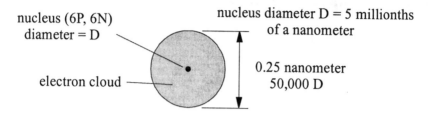

[a] Carbon atom at normal pressure
Density = 2 grams per cubic centimeter

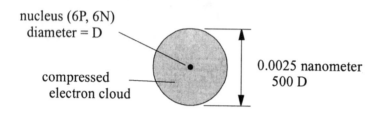

[b] Compressed to white dwarf star
Density = 2 tons per cubic centimeter

[c] Compressed to neutron star
Density = 250 million tons per cubic centimeter

Figure 2-1: Approximate dimensions of a carbon atom under conditions of [a] normal pressure, [b] white dwarf star, and [c] neutron star

Diagram (b) shows the white-dwarf state, which is the final condition for stars with masses less than 8 times that of our sun. A density of 2 tons per cubic centimeter is enormous, yet the atom still has a large amount of space left in the electron cloud circling the nucleus. The force from the electrons is sufficient to resist the enormous gravitational pressure of the very compact white dwarf star.

The *Pauli Exclusion Principle* is an important concept of quantum mechanics, which states that electrons can only occupy specific energy states. When an electron cloud has been compressed to the final white-dwarf condition, there are no spare energy states left for the electrons to occupy, and so the electron cloud cannot shrink any further.

When a star with more than 10 times the mass of our sun runs out of nuclear fuel, gravitational pressure exceeds the pressure that can be exerted by the electron cloud, and so the electron cloud is destroyed. The star collapses catastrophically, and the electrons are forced into the protons to form neutrons. With the electron cloud eliminated, the atom shrinks to the size of its nucleus. Relative to a white dwarf star, the diameter of the atom decreases by a factor of 500 and so its density increases by the cube of 500, which is 125 million. The density increases from that of a white dwarf star (2 tons per cubic centimeter) to that of a neutron star (250 million tons per cubic centimeter).

To achieve a physical feeling for the tremendous density of a neutron star, consider a fictitious exercise. With earth-moving equipment, dig a hole 1500 ft deep over an area of 50 acres, to cover a square plot 1500 ft on a side. This material is put into a super-compactor, which compresses the material into a volume of one cubic centimeter. That cube would have the density of a neutron star, about 250 million tons per cubic centimeter.

Although the density of a neutron star may seem unbelievably high, it is certainly not infinite. It is the same density that exists here on earth within the nucleus of every atom. The electrons of the atom are eliminated in a neutron star, and so the density of the star increases to the density of the atomic nucleus.

A neutron star consists entirely of tightly packed neutrons. Since there is no space left for further contraction, a neutron star has reached the greatest possible density of matter, one billion tons per teaspoon. Both Gamow and Einstein endorsed this principle, and it is solidly supported by extensive physical experiments.

The modern Big Bang theory is based on the singularity principle, which requires a density of matter that is many, many times greater than the density of a neutron star. The Einstein equations are used to justify the singularity principle, even though Einstein absolutely rejected the physical singularity.

Chapter 3

Our Mysterious Universe

The Characteristics of Our Universe

Our Milky Way Galaxy

Our sun lies within an enormous spiral galaxy, which we call the Milky Way galaxy. This is similar to the Whirlpool galaxy, which was shown in Fig. 1-1 and on the front cover. It has the shape of a disk, which is 100,000 light-years in diameter and 3500 light-years thick at the center. Our sun is located 2/3 of the distance from the center to the circumference. Our sun lies within a spiral arm and is near the center of the galaxy disk.

Our word "galaxy" was derived from the Greek word for "milk". The milk-like pathway across the sky, which we call the Milky Way, is our edge-on view of the galaxy disk. It is the light emitted by billions and billions of very distant stars that lie within our Milky Way galaxy.

Because the Milky Way galaxy is rotating, our sun is moving at a velocity of 250 kilometers per second. The galaxy makes a complete revolution in 200 million years. Our galaxy contains 100 billion stars, which have a mass equivalent to 10 billion suns.

What Is a Nebula?

Up until the early 1900's, most astronomers believed that our Milky Way galaxy comprised our complete universe. Stars look like points of light to the telescope, but our heavens also contain fuzzy extended objects that were called nebulae. Certain astronomers believed that some nebulae were very distant galaxies, like our own Milky Way galaxy, but this concept was widely disputed. The general attitude is illustrated in the following excerpt from a popular 1905 book on astronomy by Agnes Clerke, entitled *The System of the Stars* [19] (p.1):

"The question of whether nebulae are external galaxies hardly any longer needs discussion. It has been answered by the progress of research. No competent thinker, with the whole of the available evidence before him, can now, it is safe to say, maintain any singular nebula to be a star system of co-ordinate rank with the Milky Way. A practical certainty has been attained that the entire contents, stellar and nebula, of the [celestial] sphere belong to one mighty aggregation, and stand in ordered mutual relations within the limits of one all-embracing scheme."

In this quotation, Clerke was insisting that all nebulae lie within our Milky Way galaxy.

The mysterious nebulae had been observed for many years. In 1784, the locations of the larger nebulae were listed by Charles Messier (1730-1817) in his Catalogue of Nebulous Objects. This catalogue became the primary reference for locating nebulae, and still serves that purpose today. The name M51 for the Whirlpool galaxy shown in Fig 1-1 and on the front cover is the designation given by Messier.

Messier was searching for comets, and he prepared his nebulae catalogue to distinguish comets (which move) from nebulae (which are fixed). Although Messier discovered 15 comets, it is his *Catalogue of Nebulous Objects* that made the Messier name famous. Messier lost his astronomer job when the French Revolution erupted in 1789.

After Clerke's book, astronomers developed methods of measuring stellar distances, and were able to prove that some nebulae lie outside our Milky Way galaxy. Certain nebulae are stellar clusters and gaseous clouds, which lie within our Milky Way galaxy, but others (such as M51) are distant galaxies that are far beyond our Milky Way galaxy.

Distances to the Stars

The reason for this confusion is that it is extremely difficult to measure the distances of stars. Shortly after Clerke's book was published, reliable means for estimating stellar distances were achieved, and this produced a revolution in astronomy.

The most distant planet that we can see with the naked eye is Saturn, which is about 1.4 billion kilometers away. This is 3700 times further than the moon.

It is convenient to express astronomical distances in terms of the time for light to travel. It takes light 1.25 seconds to reach us from the moon, 8.3 minutes to reach us from the sun, a minimum of 66 minutes to reach us from Saturn, and a minimum of 4.0 hours to reach us from Neptune, which is the most distant planet of reasonable size.

However, planet distances are extremely small in comparison to the distances of stars. The nearest star is Proxima Centauri, which is 4.2 light-years away. Since light takes 4.2 years to reach us from this star, Proxima Centauri is 9200 times further than Neptune (4.0 light-hours), and is 100 million times further than the moon (1.25 light-seconds), which is the greatest distance traveled by man. Yet Proxima Centauri is a very close star. Our Milky Way galaxy has a diameter of 100 thousand light-years, which is 24,000 times the distance to Proxima Centauri.

Parallax Method for Measuring Stellar Distances

The measurement of astronomical distance starts with the parallax effect. As the earth revolves around the sun, the images of nearby stars shift over the year relative to the distant stars. This shift is called parallax, a process that our eyes and brain use to give us depth perception.

The images received by our two eyes are not exactly the same, and our brain compares these images to distinguish close from distant objects. To observe this effect, hold a finger in front of your nose and look with one eye at a time. As you switch from one eye to the other, the image of the finger moves back and forth relative to the distant background. The closer the finger is held to the nose, the greater is the relative motion. The reader should stop and perform this experiment in order to understand the parallax principle.

In like manner, the rotation of the earth around the sun causes the image of a nearby star to move relative to the distant stars by an amount that is inversely proportional to distance. The motion is ±1.0 arc second for a theoretical distance of 3.26 light-years. (This distance, 3.26 light-years, is called a *"parsec"*, and is the primary distance measure used by astronomers.) If a star shifts by ±0.1 arc seconds over the year, the star is 32.6 light-years away (10 parsecs).

With modern telescopes, astronomers can measure with parallax the distances to stars out to 200 light-years. The measurement is accurate to one light-year for a distance of 30 light-years, but accuracy is very poor for stars beyond 200 light-years.

There are 100 stars within 20 light-years. The volume of a sphere is proportional to the cube of its radius, and so the number of stars is approximately proportional to the cube of the distance. Consider for example the stars within 40 light-years (2 times 20 light years). Since the cube of 2 is 8, there should be about 8 times as many stars within 40 light-years as there are within 20 light-years. There are 100 stars within 20 light-years, and so there are about 800 stars within 40 light-years.

This reasoning shows that there are about 12,000 stars within 100 light-years, and 100,000 stars within 200 light-years. Since parallax can be used with moderate accuracy out to 100 light-years and with crude accuracy out to 200 light-years, astronomers can use parallax to measure distances to 12,000 stars with moderate accuracy, and the distances to 100,000 stars with crude accuracy.

With modern telescopes, particularly the Hubble space telescope, distance measurements by parallax have improved immensely.

Measuring the Distances of Remote Stars

To measure the distances of stars much further away than 100 light-years, astronomers use the spectra of the light from stars to classify stars into various types. They have found that stars of a certain spectral type radiate approximately the same power. The received power is inversely proportional to the square of the distance, and so by estimating the radiated power of a star one can estimate its distance.

An accurate yardstick for astronomy was developed by astronomer Henrietta Leavitt (1868-1921) in 1908, based on Cepheid variable stars. These are stars that vary periodically in luminosity, and are named after the star Delta Cephei. Leavitt observed the Cepheid variable stars located in the Small Magellanic Cloud, which is a cluster of stars outside our Milky Way galaxy, about 300,000 light-years from earth. Since all stars in the Small Magellanic Cloud are at approximately the same distance, she was able to compare them directly with one another.

Leavitt found that the period of variation of a Cepheid variable star is related to the light power that it radiates. By examining nearby Cepheid variable stars, the distances to which could be measured by parallax, astronomers were able to relate the period of variation of a Cepheid variable to the absolute light power that the star radiates.

Measuring the Radial Velocity of a Star

Although it is extremely difficult to measure the distance to a star, the velocity of a star in the radial direction can be measured easily and accurately. Radial velocity is the component of velocity either toward us or away from us, along a "radial" line from earth to the star.

An astronomer studies the light from a star by passing the light through an instrument that acts like a prism, which separates the wavelengths into a rainbow pattern. From this spectral pattern one can accurately measure the radial velocity of the star. It is extremely difficult to measure velocity in the transverse direction, perpendicular to the radial line to the star.

An element in the atmosphere of a star generates a unique pattern of lines in its light spectrum. These consist of bright lines that occur at wavelengths where an atom radiates energy, and dark lines that occur where an atom absorbs energy. In the light received from a star, these spectral lines are shifted in wavelength by an amount approximately proportional to the radial velocity between the star and the earth. Spectral lines of hydrogen are the most common, but lines from many other elements are also observed.

This wavelength shift is called the Doppler effect, named after the Austrian scientist, Christian Doppler (1803-1853), who discovered it in 1842. If the star is moving away from us, the spectral lines are shifted toward the red end of the spectrum, and so astronomers say that the star has *redshift*. If the star is moving toward us, the spectral lines are shifted toward the blue end of the spectrum, and so the star has *blueshift*.

The Doppler wavelength shift divided by the normal wavelength of a spectral line is approximately equal to the radial velocity of the star divided by the speed of light. By measuring the Doppler wavelength shifts in the spectrum from a star, an astronomer can accurately determine the velocity of the star toward us or away from us.

The Hubble Expansion of the Universe

One of the first astronomers to prove that some nebulae are distant galaxies was Edwin Hubble. He had the advantage of working on the new Mount Wilson telescope, which was the best telescope in the world at that time.

Hubble was able to resolve Cepheid variable stars in the galaxy M31 of the constellation Andromeda, which is 2.3 million light-years away, and in the M33 galaxy of the constellation Triangulum, which is 2.6

million light-years away. By this means he estimated the distances to M31 and M33.

About 15 years later, astronomers discovered that there are two types of variable stars: the classical Cepheid variable stars and the RR Lyrae stars. [19] (p. 35) This effect confused Hubble's measurement, and so he estimated the distances to M31 and M33 to be half of the actual distances.

Hubble examined the brightest stars in the M31 and M33 galaxies, and assumed these to be super-giant stars. He assumed that the brightest stars in more distant galaxies would radiate the same power as those in the M31 and M33 galaxies. With this approach he estimated distances to several galaxies more distant than M31 and M33, which were too far away to observe Cepheid variables.

Hubble categorized these galaxies into different types depending on their shapes. He assumed that all galaxies of a given shape have approximately the same dimensions. With this principle, he estimated the distances to galaxies much further away.

Hubble compared his distance measurements with the radial velocities of the galaxies determined from the Doppler wavelength shifts obtained from the galaxy spectra. He discovered that except for the close galaxies, M31 and M33, all galaxies have redshift and so are moving away from us. He found that the radial velocity of a galaxy is approximately proportional to its distance. This discovery was published by Hubble in 1929, and created a revolution in astronomy. It showed that our universe is expanding. This expansion of the universe is called the Hubble expansion, and the ratio of galaxy velocity divided by galaxy distance is called the *Hubble constant*.

There were many sources of error in Hubble's measurements of galaxy distance, and so his galaxy distance estimates were 8 times too short and his Hubble constant was 8 times too large. One source of error was the confusion between the two classes of variable stars, which was discussed previously. Another problem is that some of the "super-giant" stars that Hubble observed were actually regions of brightly illuminated hydrogen gas. [19] (p. 35)

Nevertheless, even though Hubble's initial measurement of the Hubble constant had serious errors, his basic discovery was correct. Our universe is indeed expanding.

Meaning of the Hubble Expansion

The meaning of the Hubble expansion of the universe can be

32   *The Mystery of Creation*

clarified by considering a rubber band that is stretched at a constant rate. At an instant of time (called t_1) mark a series of dots on the band separated from one another by 10 mm, and label these dots as follows:

```
D'  C'  B'  A   B   C   D   E
*   *   *   *   *   *   *   *
```

The distance from A to B is 10 mm, from A to C is 20 mm, from A to D is 30 mm, etc. Now look at the band later (at time t_2) when it has stretched by 10 percent, so that the distance between neighboring dots has increased to 11 mm. The distance from A to B is now 11 mm; from A to C is 22 mm; from A to D is 33 mm, etc. Between time t_1 and time t_2, point B moves 1 mm away from A, point C moves 2 mm away from A, point D moves 3 mm away from A, etc. The relative velocity between points along the band is proportional to the distance between the points. This indicates that the rubber band is being stretched at a constant rate.

We can extend this analogy to two dimensions by considering a balloon being blown up at a constant rate. Mark on the surface of the balloon an array of dots. As the balloon expands, the dots recede from one another at velocities proportional to the distances between the dots.

Modern Measurements of the Hubble Constant

Modern telescopes (particularly the Hubble Space Telescope) have allowed astronomers to make accurate measurements of the Hubble constant. Cepheid variable stars can now be observed in very distant galaxies. Another approach uses a particular type of supernova (Type 1a), which radiates a nearly constant amount of power. These supernovas radiate the power of 3 billion suns for a brief period and so can be observed in very distant galaxies.

A Type 1a supernova is formed when a white dwarf star closely orbits a red giant star, and sucks off material from the red giant star. When the mass of the white dwarf reaches a critical value, the white dwarf suffers catastrophic gravitational collapse and explodes as a Type 1a supernova.

Two teams are measuring the Hubble constant; one led by Alan Sandage and the other by Wendy Freedman. The Sandage study is based on supernovas and the Freedman study is based on Cepheid variables. The average Hubble constants recently reported by Sandage and Freedman are 18.7 and 21.5 km/sec per million light-years. ***This book assumes a Hubble constant of 20 km/sec per million light-years, which*** *is a good average of these recent measurements.*

The Hubble constant is usually expressed in terms of the parsec, which is 3.26 light-years. The Sandage and Friedman data are usually specified as 61 and 70 km/sec per megaparsec (million parsecs). [27]

The Observable Universe

Let us assume that the universe is expanding uniformly at our assumed Hubble constant of 20 km/sec per million light-years. If a galaxy is 10 million light-years away, it should recede (move away from us) at a velocity of 200 km/sec. A galaxy 100 million light-years away should recede at 2000 km/sec, and a galaxy 1000 million (or one billion) light-years away should recede as 20,000 km/sec. Therefore, a galaxy that is 15,000 million (or 15 billion) light-years away should recede at 300,000 km/sec, which is the speed of light.

If a galaxy is moving away from us at the speed of light, it cannot be seen. This argument indicates that 15 billion light-years should be our limit of observation. We presumably cannot observe a galaxy more distant than 15 billion light-years. Consequently our *observable universe* is generally considered to be a sphere with a radius of 15 billion light-years.

The Apparent Age of the Universe

An obvious explanation of the Hubble expansion of the universe is that the universe began billions of years ago as a highly dense mass that exploded with a Big Bang and has been expanding ever since. Let us assume that the universe has always expanded at the same rate, 20 km/sec per million light-years. With this assumption we can easily extrapolate the universe backward in time. Since a galaxy 15 billion light-years away would recede at the speed of light, our universe should have begun as a very dense mass 15 billion years ago. This book calls this time period, 15 billion years, the *apparent age of the universe*, which is denoted T_0. The Hubble constant (20 km/sec per million light-years) is commonly denoted H_0.

There are many versions of the Big Bang theory, but essentially all of them assume that the Hubble constant was greater in the past. Consequently, they compute a true age of the universe that is less than the apparent age. Most versions of the Big Bang theory predict a true age of the universe that lies between $(2/3)T_0$ and T_0, or between 10 and 15 billion years.

Alternatives to the Big Bang Theory

There are two major alternatives to the Big Bang theory: (1) the postulate that the Hubble expansion is only apparent, and (2) the Steady-State Universe theory. Although the original Steady State Universe theory was abandoned by its sponsors, the Yilmaz gravitational theory yields a cosmology model that satisfies the major principles of the Steady-State Universe theory but does not have the limitations of the original theory.

The Hubble Expansion Is Apparent

Some scientists argue that the universe expansion is apparent, that the universe is not actually expanding. Effects other than velocity can cause a spectral redshift. These effects may make the universe appear to expand, even though it is not.

Paul Marmet [28] has proposed a redshift effect that has a solid scientific foundation, although it does not seem to be sufficient to explain the Hubble expansion quantitatively. He has shown that a photon of light loses energy when it collides with a gas molecule, and thereby experiences a redshift. The direction of the light does not change. This redshift effect cannot be observed in the earth's atmosphere, because the density of gas is too high. It cannot be observed in the laboratory, because it requires too long a path to create a measurable effect.

However, Marmet has shown that his effect explains the variation of the redshift of radiation across the disk of the sun. The redshift of radiation from the limb of the sun is greater than that from the center. Thus Marmet has experimental evidence to support his theory.

Details of the Marmet's theory are given in *Story* [4], Appendix A. The theory indicates that every collision of a photon with a hydrogen molecule produces a redshift equivalent to a velocity change of 2 mm/sec. The number of collision is proportional to the density of the molecules and to the length of the path. The Marmet redshift would produce an apparent expansion of the universe that is consistent with recent measurements of the Hubble constant if the universe has an average gas density of 33,000 hydrogen atoms per cubic meter.

Gas densities that are very much larger than this have been observed by astronomers. In a gaseous nebula, the gas density is typically 100 billion hydrogen atoms per cubic meter.

Nevertheless, as shown in Appendix A, present estimates give an average density of matter in the universe that is very much smaller than the Marmet requirement. These estimates are based on gravitational effects associated with galaxy motions in a large galaxy cluster. Measurements of galaxy velocities within a cluster show that there must be about 325 times as much dark matter (which we cannot see) in the cluster as there is luminous matter (which we can see). This analysis yields an estimated average density of matter over the cluster equivalent to 2 hydrogen atoms per cubic meter.

The gas density required by the Marmet cosmology theory is 16,000 times greater than this estimate of 2 hydrogen atoms per cubic meter. This indicates that the Marmet redshift effect is probably not sufficient to explain the Hubble expansion of the universe.

On the other hand, we will see later in this chapter that the Marmet redshift effect gives a promising explanation for the extreme redshift of the quasar and the excess redshifts observed in certain galaxies.

The Steady-State Universe Theory

There is another well known alternative to the Big Bang theory, called the *Steady-State Universe* theory. This theory was proposed in 1948 by the noted astrophysicists Fred Hoyle (1915-2001), Hermann Bondi, and Thomas Gold. The Steady State Universe theory was seriously considered by many astronomers until it was eclipsed in the stampede toward the Big Bang theory that occurred in the late 1960's.

The Steady State Universe theory assumes that the age of the universe is infinite. The theory postulates that diffuse matter is being created throughout space to compensate for the Hubble expansion. The required creation of matter is far too small to be observed directly. Based on the present value of the Hubble constant, one hydrogen atom would be created every year within a volume of one cubic kilometer.

This theory postulates that the diffuse matter gathers into huge clouds that coalesce to form new stars and galaxies. In this manner, our universe continually changes, and thereby stays eternally young even though it is infinitely old.

What could produce this spontaneous creation of matter? The original Steady-State Universe theory did not have an answer. Is the matter created *out of nothing*? This may seem unacceptable, but should be no less acceptable than the Big Bang postulate that the whole universe was instantaneously created *out of nothing* at the Big Bang.

During World War II, George Gamow worked in developing the atomic nuclear bomb and Fred Hoyle was engaged in radar development. After the war, both scientists directed their efforts to astronomy, and wrote popular books on the subject. Gamow explained the universe expansion by postulating that our universe began billions of years ago in an enormous explosion. Hoyle presented his alternate Steady-State Universe theory. As a criticism of the Gamow concept, Hoyle used the term "Big Bang" to characterize the enormous explosion that presumably created our universe. The name stuck, and since that time the Gamow theory was called the "Big Bang theory".

Up until 1965, both the Big Bang theory and the Steady-State Universe theory were seriously considered by astronomers. Then cosmic microwave background radiation was discovered, which had been predicted by Gamow as a consequence of the Big Bang, The Steady-State Universe theory did not have an answer for this radiation, and so interest in this concept rapidly waned. The co-sponsors of the theory, Bondi and Gold, lost interest in what had become an unpopular theory, and in time Hoyle abandoned his theory. [19]

The Big Bang Theory

Let us return to the Big Bang theory. It represents two radically different cosmological theories: (1) the Gamow Big Bang theory which postulated that the initial universe had the density of a neutron star, and (2) the modern Big Bang theory which claims that the universe began as a singularity. With the Gamow Big Bang postulate, the observable universe would have initially been about the size of the orbit of Mars around the sun, whereas the modern Big Bang theory predicts it would have been smaller than a dime, and probably microscopic in size.

Definition of the Singularity

Let us examine in more detail the meaning of the singularity. This concept was expressed as follows in a book by Filkin [17] (p. 104), titled, *Stephen Hawking's Universe*:

> *"Stephen [Hawking] and Roger Penrose published a paper in 1970 which proved that, if Einstein's mathematics were correct, a singularity had to result from a black hole, and had to exist at the start of the universe. - - - The paper argued that if relativity as*

explained by Einstein is correct — and all of the evidence from observation seems to keep confirming it — then the universe must have started with a big bang explosion out of a singularity. The equations do not allow an alternative."

Filkin directed a 1997 television series for the Public Broadcasting System, also titled, *Stephen Hawking's Universe*. This book was a supplement to that television series.

What is meant by a "singularity"? Joseph Silk is the author of three books on the Big Bang theory. One of these [21] was officially endorsed by the *Scientific American*, and is a volume in the *Scientific American Library*. This book [21] (p. 66) gives the following definition of a singularity, as recognized by Big Bang cosmologists:

Singularities: *The universe began in time zero in a state of infinite density. At least the existence of such a singular state is the expectation from extrapolating the present universe backward in time. Of course, the phrase 'a state of infinite density' is completely unacceptable as a physical description of the universe. An infinitely dense universe would be what is called a 'singularity', where the laws of physics, and even space and time, break down. The resolution of the paradox is that our theory of gravitation has broken down before reaching this extreme state. The first 10^{-43} sec is inaccessible in our current theories. The end of the period, called the Planck instant, represents the beginning of time according to those theories. Nevertheless, to the extent that one accepts Einstein's theory of gravity, singularities are predicted to exist in nature."*

Silk [21] (p. 79) illustrates two versions of the modern Big Bang theory, which start 10^{-35} sec after the Big Bang. One of these applies the "inflation" postulate. Silk explains that most of the recent Big Bang models have endorsed the "inflation" concept that was proposed by Alan Guth in 1981. This concept postulates that in the initial period the universe size doubled every 10^{-35} second, until it reached a much larger size. Then the rate of expansion slowed.

Silk [21] illustrates on page 79 the early (pre-inflationary) Big Bang concept and one that applies the inflation postulate. Both of these started 10^{-35} sec after the Big Bang. In the early Big Bang model, the observable universe began as a body that was 1 mm in diameter. In the inflationary Big Bang model, the observable universe began with a microscopic diameter of 10^{-24} mm.

As shown in Chapter 1, Dickinson [24] (p. 118) gives a pictorial representation of the Big Bang singularity that includes inflation. He states that during the inflation period, the observable universe "expands from one trillionth the size of a proton to the size of a baseball".

These singularity predictions of modern cosmology are so extreme some readers may feel that the author must be exaggerating. The best supporting evidence that the author can find is this book by Dickinson [24]. It is a beautifully illustrated third edition of a classic book on astronomy. It is obvious that the concepts presented in the book have wide support in the astronomical community. Page 118 of Dickinson's book gives a dramatic graphic presentation of the Big Bang singularity, which began with a body that was, "one trillionth the size of a proton".

Cosmic Microwave Background Radiation

In the late 1960's, there was a stampede of astronomers to the Big Bang theory after cosmic microwave radiation was discovered. Big Bang proponents claimed that cosmic microwave background radiation was the "smoking gun" that proved the validity of the Big Bang theory. However, the primary cause for the stampede to the Big Bang was the intensive research effort involved in computer studies of the Einstein equations, which began about the same time. As we saw earlier, these studies led invariably to the Big Bang theory.

Although cosmic microwave radiation was originally predicted by Gamow as part of his Big Bang theory, this stampede to the Big Bang theory in the mid 1960's discarded the Gamow Big Bang theory, and replaced it with the modern Big Bang theory with its singularity concept.

Let us examine cosmic microwave background radiation, which was a key milestone in the development of the Big Bang theory. In 1965, Arno Penzias and Robert Wilson, two physicists working at Bell Laboratories were performing measurements on a sensitive microwave antenna that had been developed for satellite communication. When communication satellites generated greater power, there was no longer a communication need for this sensitive instrument, and so the antenna was applied to basic research.

Penzias and Wilson discovered spurious microwave signals in their antenna, which they could not explain. These signals represented cosmic radiation coming from all directions in space. Gamow had predicted that the expansion of the universe should have caused optical radiation (emitted by the early universe soon after the Big Bang) to be reduced in frequency, and should be observable today at microwave frequencies. He predicted cosmic microwave radiation coming from all directions that is equivalent to the radiation from an ideal blackbody. To understand this radiation, let us examine the spectrum of the radiation from an ideal blackbody.

Spectrum of Radiation from an Ideal Blackbody

Astronomers frequently use the spectral characteristics of a blackbody in their studies, because the radiation from a star approximates that from an ideal blackbody. An object at room temperature continually radiates energy in terms of heat, and it absorbs heat from the environment. Heat radiation is like light, except that it has a longer wavelength. The blacker the surface of an object, the better the object absorbs radiation, and the better it radiates. A *blackbody* is a physical idealization that radiates the maximum possible energy from a body at a particular temperature.

One can build a nearly ideal *blackbody* by machining a spherical cavity inside a block of metal, and cutting a small hole into the cavity. Inside the cavity, radiation is emitted from each portion of the spherical surface, and continually reflects off other surfaces, until a small amount leaks out of the hole. The energy escaping from the hole is close to ideal *blackbody* radiation. A *blackbody* is an idealized physical concept, yet it can be closely approximated by physical equipment.

The general spectrum of *blackbody* radiation is given in Fig. 3-1, which shows the spectrum versus frequency. The spectral plot versus wavelength has a somewhat different shape. The frequency scale is expressed in terms of the half-power frequency f_h. Half of the power falls at frequencies less than f_h and half falls at higher frequencies. Area under the curve is proportional to power, and so the areas under the plot on each side of the vertical line are equal.

It is convenient to express the half-power frequency f_h in terms of the equivalent half-power wavelength λ_h. The Greek letter λ (lambda) is used to denote wavelength. Since λ is the Greek equivalent of the Roman letter l, the symbol λ is convenient for denoting *waveLength*. The half power wavelength λ_h is equal to the ratio c/f_h, which is the speed of light c divided by the half-power frequency f_h.

The temperature of a blackbody determines the actual frequencies that it radiates, but the shape of the spectrum is the same for all temperatures. The half-power wavelength λ_h is related as follows to the blackbody temperature T, expressed in degrees Kelvin (°K):

λ_h = 4.107/T millimeter (mm)

Kelvin temperature is measured above absolute zero temperature, which is -273 degrees Celsius. At absolute zero temperature, the random motion of molecules is zero. The temperature in degrees Kelvin is obtained by adding 273 degrees to the temperature in degrees Celsius.

The light radiated from our sun has a spectrum approximating that of an ideal blackbody at a temperature T of 5770 °K. The corresponding value for the wavelength λ_h computed from the above formula is 0.000712 mm, or 0.712 micrometers (millionths of a meter).

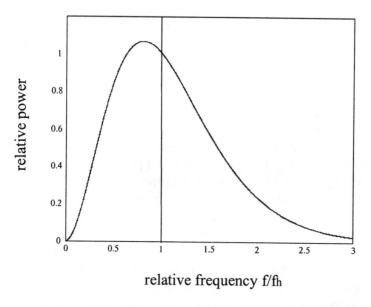

Figure 3-1: Spectrum of a blackbody radiator

The intensity of the radiation from a blackbody is related as follows to the temperature T of the body, expressed in degrees Kelvin:

P/A = 5.68 (T/1000)^4 watt/cm^2

Intensity is the power P radiated per unit of surface area A. For our sun, with a blackbody temperature of 5770 °K, this formula gives an intensity of 6300 watts per square centimeter at the surface of the sun.

The radiation from most stars approximates that of a blackbody.

Consequently, from the general spectral plot of Fig 3-1 one can find from the spectrum of a star its half-power wavelength λ_h. From this one can calculate with the above equations the blackbody temperature T and the power radiated from the star per unit of surface area (the radiation intensity). If the distance to the star is known, one can determine the power radiated from the star. Dividing the total power by the radiation intensity gives the surface area of the star, and from this one can obtain the star diameter.

Dickinson [24] (pps. 80-81) discusses the use of the spectral characteristics of stars to estimate their stellar characteristics.

Blackbody Temperature of Cosmic Microwave Radiation

Gamow made several estimates of the blackbody temperature of his cosmic microwave radiation, which varied from 5 degrees Kelvin to 20 degrees Kelvin. In the early 1960's, Prof. Robert Dicke of Princeton University was studying the Big Bang theory. His graduate student, P. James E. Peebles, investigated Gamow's prediction of cosmic microwave radiation, and considered building an antenna to measure this radiation. Peebles estimated that the blackbody temperature should be 30 degrees Kelvin. Then he discovered that Penzias and Wilson had already measured similar radiation in their antenna. The antenna had detected unexplained electrical disturbance signals with a spectrum that corresponded to a blackbody temperature of 3.5 degrees Kelvin.

This discovery of the cosmic microwave background radiation was hailed by Big Bang theorists. Since this cosmic radiation had been predicted by the Big Bang theory, it was proclaimed to be proof that the Big Bang theory was correct. However, the Big Bang proponents fail to mention that this radiation was only predicted in a qualitative sense.

An equation given above shows that the intensity of radiation from a blackbody varies as the fourth power of temperature. Consequently, the 30 degree Kelvin estimate by Peebles corresponds to a radiation intensity that is 5000 times greater than the intensity for a blackbody at 3.5 degrees Kelvin.

In 1948, Ralph Alpher and Robert Herman, two graduate students working with George Gamow, had published a paper in the scientific journal *Nature* that predicted cosmic microwave radiation corresponding to a blackbody temperature of 5 degrees Kelvin. Since this is the closest estimate to the measured temperature, it is the value generally quoted by Big Bang proponents.

Much more accurate measurements of this cosmic radiation were obtained from the Cosmic Background Explorer (COBE) satellite in 1989. This satellite measured cosmic microwave radiation coming uniformly from all directions that corresponded very accurately, in intensity as well as spectrum, to the radiation from an ideal blackbody at a temperature of 2.73 degrees Kelvin. For a blackbody temperature T of 2.73 degrees Kelvin, the formula shown previously gives a half-power wavelength of 1.50 millimeters, which lies in the high microwave region.

The Big Bang theory claims that this cosmic microwave background radiation is the cooled relic of optical radiation that was emitted from the early universe about 300,000 years after the Big Bang. However, an important weakness of this assumption is that the cosmic radiation emanates with extreme uniformity from all directions. In the data obtained from the COBE satellite, the energy of radiation received from different directions varies by only a few parts in 100,000.

This implies extreme uniformity of the universe at this early period. How did this highly uniform early universe create what we observe today? Not only is our universe separated into galaxies, but the galaxies are not spaced evenly.

Cosmic Radiation for the Yilmaz Cosmology Model

The original Steady-State Universe theory could not explain the cosmic microwave background radiation, and so the theory was eventually abandoned even by its sponsors. The Yilmaz cosmology model is similar to the Steady-State Universe theory. However it differs from the Steady-State Universe theory in many important respects, because it is based on the Yilmaz gravitational theory rather than on the Einstein General Relativity theory.

As shown in Appendix B, the Yilmaz cosmology model directly predicts cosmic microwave background radiation, and does so much more accurately than the Big Bang theory. The predicted radiation is equivalent to the emission from an ideal blackbody at a temperature between 2.1 and 3.4 degree Kelvin. This agrees closely with the 2.73 degree Kelvin blackbody temperature measured by the COBE satellite. According to the Yilmaz theory, this cosmic microwave radiation should be emitted with high uniformity from all directions, which agrees with the COBE measurements.

The Black Hole

We saw in Chapter 1 that the *modern Big Bang theory* is based on the *singularity* principle. Closely related to the Big Bang singularity is the black hole singularity. Both of these singularity concepts evolved from Einstein's General theory of Relativity.

The first claim that the Einstein theory predicted a singularity occurred in a 1939 scientific paper by Robert Oppenheimer (who later managed the Manhattan atomic bomb project) and his graduate student, Hartland Snyder. They applied the Einstein equations to a massive, highly dense star, and concluded that the star must contract "indefinitely" to form a singularity. This analysis was related to the famous solution of the Einstein equations by Karl Schwartzschild, and so the predicted singularity was called a "Schwartzschild singularity". The Oppenheimer-Snyder article [11] concluded with:

"When all thermonuclear sources of energy are exhausted, a sufficiently heavy star will collapse. Unless fission due to rotation, the radiation of mass, or the blowing off of mass by radiation, reduce the star's mass to the order of that of the sun, this contraction will continue indefinitely."

The next month, Einstein flatly rejected this concept with an extensive analysis in *Annals of Mathematics* [12] (p. 936), which was summarized with:

*"The essential result of this investigation is a clear understanding as to why the '**Schwartzschild singularities**' **do not exist in physical reality**. . . . The 'Schwartzschild singularity' does not appear for the reason that matter cannot be concentrated arbitrarily. And this is due to the fact that otherwise the constituting particles would reach the velocity of light."*

A few years later, Oppenheimer became the manager of the Manhattan atomic bomb project. He never pursued this issue further, and neither did any other scientist during Einstein's lifetime. The collapsing star predicted in the Oppenheimer-Snyder paper later became known as a "black hole". The black hole theory requires that the star inside the black hole must collapse to form a singularity having a diameter of zero.

Einstein never accepted the black hole singularity or the Big Bang singularity. After Einstein's 1939 rejection of the black hole singularity, no scientist claimed that the Einstein theory predicted a singularity during Einstein's lifetime.

The Schwartzschild Solution to General Relativity

When Einstein presented the equations for his General theory of Relativity, he had only been able to derive approximate solutions to his theory. Karl Schwartzschild was cooperating with Einstein and was able to derive an exact solution for a very simple model, which assumed a star with a constant density of matter and no viscosity. Einstein published the Schwartzschild solution along with his general theory in 1916. The Schwartzschild solution formed the basis for devising experimental tests to verify the Einstein General theory of Relativity.

Unfortunately, Karl Schwartzschild died suddenly from the rare skin disease pemphigus even before his famous solution was printed. He had been a German army officer on the Russian front during World War I.

The Schwartzschild analysis predicted that a gravitational field should cause the speed of light to decrease. The gravitational field of a star is proportional to the ratio of its mass M divided by its radius r. In the Schwartzschild analysis, the speed of light goes to zero when the mass-to-radius ratio of a star is 240,000 times greater than the value at the surface of our sun. This condition where the speed of light goes to zero is called the "Schwartzschild limit", because the Schwartzschild analysis does not yield an answer for larger gravitational fields.

Einstein was not concerned about the Schwartzschild limit, because this limit occurs at a mass-to-radius ratio that is about one quarter million times the maximum value occurring within our solar system. Einstein was interested in applying his theory to practical situations, and so he ignored this limit to the Schwartzschild analysis.

The Black Hole Prediction

In their 1939 paper, Oppenheimer and Snyder found that they could achieve a solution to the Einstein equations for a mass-to-radius ratio exceeding the Schwartzschild limit if they assumed that the diameter of the star decreased with time. Consequently, they concluded that if the mass-to-radius ratio exceeds the Schwartzschild limit, the star must contract "indefinitely". Since there is no theoretical limit to this contraction, they concluded that the star should shrink to become a physical singularity having zero size. The mass does not change as the star shrinks, and so the density of matter within the star should become infinite as the star diameter shrinks to zero.

Einstein recognized that a physical singularity was impossible and absolutely rejected the singularity prediction of the Oppenheimer-Snyder paper. Einstein never accepted the physical singularity.

The Oppenheimer-Snyder analysis predicted that light cannot escape from a star for which the mass-to-radius ratio exceeds the Schwartzschild limit, and so such a star became known as a "black hole". During Einstein's lifetime no scientist officially pursued the Oppenheimer-Snyder result, but the "black hole" became a popular concept for science fiction. There were many accounts of the process of "falling into a black hole".

The theory indicates that if the mass-to-radius ratio exceeds the Schwartzschild limit, the star is surrounded by a spherical surface, called an "event horizon", over which the speed of light is zero. The mass-to-radius ratio at this surface is equal to the Schwartzschild limit. Light theoretically cannot escape from within the event horizon surface.

Although the black hole concept is well known to the general public, it is not generally recognized that the star that lies within a black-hole event horizon must shrink until it becomes a singularity having zero diameter and an infinite density of matter.

Modern Acceptance of the Black Hole Singularity

About a decade after Einstein's death, computers became widely available, and many scientists began to apply them to the Einstein theory. These computer studies seemed to prove that Oppenheimer and Snyder had been right. Contrary to Einstein's objection, the computer studies concluded that the "Schwartzschild singularity", which is associated with the "black hole", is a required prediction of the Einstein gravitational field equation. Therefore the black hole concept became widely accepted.

Several astronomers have made observations that are considered to be proof that black holes actually exist. But how does one prove the existence of a black hole? A black hole does not emit any radiation to indicate its presence. Astronomers are observing the gravitational effects of massive, compact bodies. According to the Einstein theory, such bodies should theoretically be black holes. However, the Yilmaz theory predicts that these bodies are actually massive neutron stars.

The Quasar

Another confusing aspect of modern astronomy is the quasar, which was discovered in 1963. Like a normal star, this object appears as a point of light to a telescope. However, its spectrum has an extremely large redshift. This redshift is generally interpreted to be a Doppler effect due to velocity. This implies that quasars are receding at velocities approaching the speed of light, and so should be billions of light-years away. At such a vast distance a typical quasar must radiate the power of 1000 billion suns for it to display its observe brightness.

Another strange aspect of the quasar is that the brightness of a quasar often varies rapidly. Many quasars vary about two-to-one in brightness over periods of months, weeks, days and even hours. This indicates that quasars must be small. If quasar brightness varies by two-to-one within one day, 50 percent of the quasar power must be radiated from a volume that is no more than one light-day thick. How can such enormous quasar power be radiated from such a small volume?

The first quasars discovered were radio sources, and so were named *quasi-stellar radio objects*, and given the acronym *quasar*. Later it was found that many of these strange star-like objects did not emit radio waves, and so the name was changed to *quasi-stellar object*, but the acronym *quasar* was retained. They are also called *QSO's*.

Quasar Observations by Astronomer Halton Arp

Because of the unbelievable properties attributed to quasars, the noted astronomer, Halton Arp, suspected that they might be much closer than was being assumed. He started making observations of quasars to obtain direct astronomical estimates of their distances.

Arp found many quasars with images very close to galaxies having much smaller redshifts. If the image of one quasar is very close to that of a galaxy, this might be a chance relationship. The quasar might be billions of light-years beyond the galaxy. However, if two or more quasars appear to be close to a galaxy, the probability of a chance relationship is remote. We ask, "What is the probability that the images of two or more quasars would fall this close to an arbitrary direction in space?" Probability considerations indicate that it is highly unlikely that the quasars are not physically close to the associated galaxy.

Arp found three quasars in the outer fringe of galaxy NGC 3842. It is theoretically possible that this is a chance relationship; that the quasars actually lay far beyond the galaxy. However, the probability that three images of extremely distant quasars would fall this close to an arbitrary direction in space is less than one in a million.

Arp found several cases like this. The possibility is essentially zero that all of these observations could be accidental, which relate quasars to galaxies of much smaller redshifts.

Halton Arp also found many cases of filament structures that directly connect quasars to galaxies having much smaller redshifts. These observations suggest that the quasar was ejected from the associated galaxy by a supernova explosion. There are several examples of a filament connecting a quasar to a galaxy, and another opposing filament on the other side of the galaxy. This suggests that the opposing filament is the reaction from a supernova explosion that ejected the quasar from the galaxy. A detailed discussion of these findings by Arp is given in *Story* [4], Chapter 11.

The quasar observations by Arp were opposed by the astronomical community, because they did not agree with the accepted dogma that quasars are billions of light-years away. He found it very difficult to get his findings published. Some journals rejected his work, and papers were often held up for years by referees. Finally in 1984, the committee that controls observation time at Palomar and Mount Wilson Observatories refused to allow Arp to use these facilities.

Halton Arp had performed distinguished research at Palomar and Mount Wilson Observatories since he received his PhD degree in 1953. He was president of the Astronomical Society of the Pacific from 1980 to 1983, and received awards from the American Astronomical Society, the American Association for the Advancement of Science, and the Alexander von Humbolt Senior Scientist Award. After being denied research facilities in California, he was forced to move to Germany to continue his career, where he joined the Max Planck Institute for Physics and Astrophysics in Munich.

Arp presented his quasar observations up to the time of his expulsion in his 1987 book, *Quasars, Redshifts, and Controversies* [14]. In his later 1998 book, *Seeing Red* [15], he included the extensive observations on quasars that he has made since he was forced to move to Germany.

Arp's books give overwhelming evidence that quasars are very much closer than is generally assumed. Nevertheless, officials in astronomy completely ignore this evidence, and continue to insist that the conventional quasar dogma is correct.

The traumatic experience of Halton Arp was summarized as follows by Fred Hoyle, Geoffrey Burbidge, and Jayant V. Narlikar [19] in their book published in the year 2000, *A Different Approach to Cosmology*:

"Arp's own colleagues at the Mount Wilson and Palomar Observatories . . . recommended to the directors of the two observatories that his observational program should be stopped, i.e., that he should not be given observing time on the [telescopes in

these observatories] to carry on with this program. Despite his protests, the recommendation was implemented, and after his appeals to the trustees of the Carnegie Institution were turned down, he took early retirement and moved to Germany where he now resides, working at the Max-Planck-Institut fur Physik und Astrophysik in Munich. . . . Thus Arp was the subject of one of the most clear cut and successful attempts in modern times to block research which it was felt, correctly, would be revolutionary in its impact if it were to succeed."

Explanations for Intrinsic Redshift

Halton Arp's astronomical observations have provided strong evidence that, for at least some quasars, the redshift of a quasar does not specify its velocity or its distance. In Arp's terminology, a quasar has an *"intrinsic redshift"* that is unrelated to its velocity.

Arp has also found evidence of galaxies with intrinsic redshift. He has photographed many examples of a galaxy directly connected by a filament to a larger galaxy having much lower redshift. The evidence suggests that the smaller galaxy was ejected from the larger galaxy. In *Seeing Red* [15], Arp gives several examples of galaxies with intrinsic redshift

What is causing intrinsic redshift? We have found the following two effects that are strong candidates for explaining intrinsic quasar redshift: *(1) gravitational redshift, to be discussed in Chapter 9; and (2) the redshift effect of Paul Marmet, discussed earlier in this chapter.* The Marmet effect can also explain intrinsic galaxy redshift.

Gravitational Redshift

When the extreme redshift of the quasar was discovered in 1963, it was recognized by astronomers that the quasar would be a much more reasonable star if the extreme wavelength shift could be attributed to an intense gravitational field rather than to an extreme velocity. Chapter 9 will show that the Einstein theory predicts that gravity produces a redshift, but cannot predict a gravitational redshift greater than 2. This did not initially rule out gravitational redshift, because the early quasars had redshifts less than 2. However, Chandrasekhar [29] proved from an analysis of the Einstein theory that a star should exhibit strong radial oscillations unless its gravitational redshift was much less than 2.

A second argument against gravitational redshift is that quasar spectra display "forbidden" spectral lines of oxygen and neon. Forbidden spectral lines are never encountered on earth and, except for quasars, are only observed in gaseous nebulae. A gaseous nebula is a large thin cloud of gas that is heated to a temperature of about 10,000 °C by the radiation from nearby stars. It was concluded that a star that has sufficient density to exhibit a large gravitational redshift would be too dense to emit forbidden spectral lines.

With these two arguments, the gravitational redshift explanation for the quasar spectrum was soundly rejected in the early quasar studies. Nevertheless, a fresh examination of the evidence shows that gravitational redshift remains a possible explanation for quasar redshift.

The argument by Chandrasekhar [29] is easily refuted. The oscillation that he derived applies only to the Einstein theory. The Yilmaz theory does not exhibit this problem. The assumption that the forbidden lines in the quasar spectra are produced by huge gaseous nebulae is inconsistent with the rapid brightness variation of many quasars. Such quasars would be too compact to contain gaseous nebulae that are sufficiently large to emit the observed lines. Hence it is not clear what is producing the forbidden spectral lines.

The Marmet Redshift

The Marmet redshift, described earlier, predicts that a cloud of hydrogen gas produces a redshift that is proportional to gas density and path length. Many gaseous nebulae have gas densities of 100,000 molecules per cubic centimeter. A cloud with this density would produce an intrinsic redshift of 0.2 for 1000 light-years of cloud thickness.

This redshift effect appears to be sufficient to explain intrinsic redshift of all quasars as well as the intrinsic redshifts observed in certain galaxies. When astronomers examine the evidence on quasars objectively, and include the overwhelming evidence provided by Arp, they will probably discover the true nature of the quasar.

The Mysterious Dark Matter

Scientific American, March 2003 (pp. 50-55), described a frantic search for mysterious "dark matter" that presumably comprises most of the universe mass. Appendix A shows that measurements of the motions of galaxy clusters indicate that there must be about 325 times as much dark matter in the cluster (which we cannot see) as there is luminous matter (which we can see). ***The obvious explanation is that this dark matter is mostly hydrogen gas.***

As shown by Silk [21] (p. 163), astronomers have rejected this obvious explanation, based on their studies of quasar spectra. They see little evidence of hydrogen lines in the spectra of quasars. Since they believe that quasars are at enormous distances, they conclude that the density of intergalactic hydrogen must be extremely low. However, Arp has proven that quasars are very much closer than astronomers are assuming. Besides, most of the intergalactic hydrogen may be molecular hydrogen (H_2) rather than atomic hydrogen (H), and molecular hydrogen is very difficult to detect. Consequently these measurements of the densities of intergalactic hydrogen are not meaningful.

This search for exotic dark matter is misguided. If astronomers would pay attention to the overwhelming quasar evidence provided by Halton Arp, their research would be more productive. They would realize that their exotic dark matter is probably nothing more than hydrogen gas.

Understanding the Einstein and Yilmaz Theories

This discussion tells us that in order to explore the mysteries of cosmology we must first understand the Einstein General theory of Relativity and the Yilmaz gravitational theory, which is a refinement of the Einstein theory. The Einstein and Yilmaz theories are theories of gravity. As a preliminary step in investigating these theories, we examine in Chapter 4 the gravitational theory of Isaac Newton.

Chapter 4

Newton's Theory of Gravity

The motions of objects here on earth and the motions of bodies throughout our universe are described with high accuracy by the theory of gravity presented in 1687 by Isaac Newton (1642-1727). Newton's theory of gravity was based on revolutionary discoveries that were made over the preceding century and a half by Copernicus, Kepler, and Galileo.

The Copernicus Revolution in Astronomy

It seemed obvious to the ancients that the sun, moon, and stars revolve around the earth every day. As the sun and moon revolve around the earth, they move a bit each day relative to the stars. There are some stars that also move relative to the fixed stars, and so were called "planets" by the Greeks, which is the Greek name for "wanderer".

The Astronomical System of Ptolemy

The Greek astronomer Ptolemy of Alexandria (c. 100-170 AD) developed elaborate formulas to describe the motions of the planets, sun, and moon, relative to the fixed stars, which were based on centuries of Babylonian astronomical measurements. This information was published by Ptolemy in his *Almagest* about AD 140. The resultant Ptolemy system ruled the field of astronomy for 1500 years.

The Astronomical Theory of Copernicus

In 1543 the Ptolemy system was challenged by the Polish astronomer Nicholaus Copernicus (1473-1543), whose Polish name was Mikolai Kopernik. Copernicus was a clergyman (possibly a priest), secretary, and physician in the Frauenburg cathedral in Polish East Prussia. Copernicus concluded that the daily motions of the sun, moon, and stars could be explained by assuming that the earth is spinning, and he discovered that the complicated motions of the planets in the Ptolemy system become simple by assuming that the earth and the planets revolve in circular orbits around the sun.

Copernicus published a preliminary version of his theory, which received harsh criticism from many sources, including the Protestant leaders Luther and Calvin, but Pope Leo X expressed open-minded interest. The concept that our mother earth was not the center of the universe contradicted the religious beliefs of many people. Besides, how could we be spinning at a fantastic speed along with our earth and not feel the motion?

The full manuscript describing the Copernicus theory was completed in 1530, but was not published until 1543, just before Copernicus died. Copernicus withheld publication of his manuscript because he was afraid of the opposition it would receive. He agreed to let friends publish it when he knew he was close to death.

The Planetary Laws of Kepler

The Copernicus system was greatly refined in the early 1600's by the German astronomer Johannes Kepler (1571-1630). In 1601 Kepler became director of Tycho Brahe's observatory near Prague, after Brahe died. For many years, Tycho Brahe (1546-1601), a Danish astronomer, had made accurate measurements of planet and star locations. With very large instruments having no optical magnification, Brahe made astronomical measurements to a precision of 10 angular seconds.

Kepler performed extensive mathematical calculations over many years to apply Brahe's data to the Copernicus model. He discovered that the orbits of the planets around the sun are ellipses, rather than circles. From his calculations, Kepler derived three accurate relations for the orbits of the planets, which are known as Kepler's Laws. These are:

(1) A planet orbits the sun in an elliptical orbit, with the sun at one focus of the ellipse.

(2) A planet moves more rapidly when nearer the sun than when further away, such that a radius drawn from the planet to the sun sweeps over an equal area for an equal time interval.

(3) The expression r^3/T^2 is the same for all planetary orbits, where r is the mean distance of the orbit from the sun, and T is the period of the orbit.

Kepler published the first two laws in 1609. He published all three laws in 1621 in his *Epitome of Copernican Astronomy*, which was Kepler's major publication. It was the first astronomical textbook based on the Copernican system, and was the primary source of information on the subject for 30 years.

Kepler died in poverty in 1630. Because of war, Kepler was unable to collect the arrears of his Imperial salary. Another personal blow to Kepler was that his mother was imprisoned 13 months for witchcraft, and died in 1622 soon after her release. [31]

Galileo and His Telescope

In 1609, Galileo Galilei (1564-1642), an Italian professor at the University of Padua, near Venice, learned of the telescope that had been invented in the Netherlands, and began making his own telescopes. In 1610 Galileo pointed his telescope toward the heavens and made observations that revolutionized astronomy.

For hundreds of years optical lenses had been made to correct visual defects. About 1608 a Dutch spectacle maker (Hans Lippershay or Jacob Metius) built the first telescope, which had a magnification of 3 to 4. It consisted of a convex objective lens and a concave eyepiece. (A convex lens is curved outward, whereas a concave lens is curved inward, like a *cave.*)

Galileo heard of this and began grinding lenses to make better telescopes. In 1609 he supplied the governor *(doge)* of Venice with an 8-power telescope. This was so valuable for naval use that Galileo's salary was doubled, and he received lifelong tenure as a professor. Later in that year, Galileo built a 20-power telescope, which he directed at the heavens.

In 1610 Galileo revolutionized astronomy with his telescope. He studied our moon and found that it has mountains and valleys like the earth. He looked at Jupiter and made the fantastic discovery that Jupiter has moons of its own. He saw that Venus has phases like our moon, which showed that Venus must revolve around the sun. Galileo published his findings, and claimed that his astronomical observations proved that the Copernicus theory must be correct.

Galileo received strong criticism for his support of the Copernicus theory from a number of influential university professors and clergymen. Galileo contributed to this conflict by writing material that was taken as a personal affront by politically powerful intellectuals. They finally convinced the Roman Catholic Church to bring Galileo to trial. In 1633 Galileo was convicted of heresy and imprisoned.

However Galileo was soon placed under house arrest, and lived comfortably in his villa near Florence. He was able to receive visitors and teach students. He wrote defiant books that were smuggled out to foreign publishers. He died in 1642. [31, 33]

The Galileo incident is often considered to be part of a continual conflict between science and Christianity, particularly the Roman Catholic Church, but this concept is simplistic. Galileo's book *Dialogue*, published in 1632, infuriated powerful intellectual leaders. These intellectuals used their political power to have Galileo's voice suppressed by the government. In Florence, Italy at the time of Galileo, the Catholic Church was the government.

Between 1450 and 1700, thousands of people were executed throughout Europe for witchcraft, in both Protestant and Catholic areas. When compared with this, Galileo's punishment was not as harsh as it may seem today.

In addition to his revolutionary astronomical studies, Galileo made great advances in our understanding of mechanics by studying the motions of falling bodies. He was motivated in this research in order to explain how our earth could be spinning so rapidly without our feeling the motion. Let us examine the research by Galileo on falling bodies.

The Motion of a Falling Body

Acceleration is the rate of change of velocity. When an object falls, it drops with a constant acceleration, called the acceleration of gravity, provided that resistance from air is negligible. This means that the velocity of descent is proportional to time.

The acceleration of gravity, denoted g, is 9.8 meters/second per second. To simplify our calculations, we round off the acceleration of gravity g to 10 meter/sec per second. Figure 4-1 shows the velocity and distance dropped of a falling body.

Diagram (a) shows that the rate of change of velocity (the slope of the velocity plot) is constant. This means that the acceleration is constant. Since the acceleration is 10 meter/sec per second, over every second the velocity increases by 10 meter/sec. The velocity is 10 meter/sec in one second, 20 meter/sec in 2 seconds, etc.

[a]

[b]

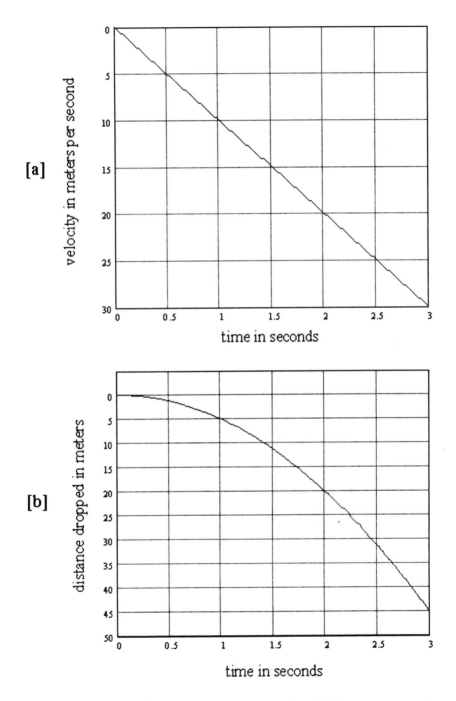

Figure 4-1: Velocity in meters per second and distance dropped in meters for a falling body, versus time in seconds

The distance dropped is proportional to the square of the time in seconds. In 3 seconds the body drops 45 meters (148 ft.). At that time the velocity is 30 meter/sec or 67 miles per hour.

Velocity is the slope of the distance plot. The slope of the distance plot gets steeper and steeper with time, which indicates that the velocity is increasing. Galileo made measurements of the motions of falling objects, which showed that they fall with a constant acceleration.

Newton's Laws of Mechanics

The theory of gravity developed by Isaac Newton (1642-1727) was based on the findings of Kepler and Galileo. The key to Newton's discovery with his invention of calculus. This gave Newton the mathematics that allowed him to express the motions of bodies in terms of general mechanical laws.

Newton learned from Galileo that bodies fall with a constant acceleration. He obtained from Kepler precise laws describing the motions of planets around the sun, and he knew the motion of the moon around the earth. By applying calculus to this information, Newton derived the following two laws of mechanics:

(1) Law of gravitational attraction: Two bodies are attracted together with a force that is proportional to the product of their masses, and is inversely proportional to the square of the distance between their centers of gravity.

(2) Law of motion: The force applied to a body is equal to the mass of the body times its acceleration, where acceleration is the rate-of-change of velocity.

Newton also gave the following corollaries of his two basic laws:

(3) A body at rest, or moving at constant velocity, stays in that condition unless a force is applied to it.

(4) For every action there is an equal and opposite reaction.

By applying calculus to these laws, Newton was able to calculate accurately the motions of bodies in our solar system. Newton's theory was published in 1687 by the Royal Society of England as *Philosophiae Naturalis Principia Mathematica*. Latin was used because scientists in all countries could read Latin. The famous scientist, Edmund Halley (1657-1742), personally paid for the printing.

Newton's invention of calculus was the key that allowed Newton to develop his physical laws. Nevertheless, he did not use calculus in his *Principia*, because other scientists did not understand it. Instead, Newton applied graphical constructions that achieved the effects of calculus.

It is commonly believed that Newton discovered the principle of gravitational attraction, but this is not true. Forty-two years before Newton published his *Principia*, Ismaelis Bouillard had postulated that mutual attraction of the planets varies inversely as the square of the distance between them. [32] It is probable that even before that both Galileo and Kepler understood that gravitational attraction holds the planets in their orbits around the sun, and holds the moon in its orbit around the earth.

What Newton actually developed was a precise set of mathematical laws, with the proof that these laws accurately describe the orbits of bodies in our solar system.

Application of Newton's Laws

Engineering Use of Newton's Laws

By demonstrating that his laws accurately explain the motions of the planets and our moon, Newton established the validity of his laws. This proved that his laws also apply to the motions of objects here on earth, and so his laws became powerful engineering tools for practical applications. In the 1600's, advances in mechanical equipment had stimulated the search for basic scientific knowledge. Engineers needed to know how to build better mechanisms, and Newton's laws had immediate practical use in engineering applications.

The famous and brilliant German mathematician, philosopher, and statesman Gottfried Wilhelm Leibniz (1646-1716) invented calculus independently of Newton. He discovered calculus in 1675, nine years after Newton, but he published it in 1684 before Newton did. With the help of the Swiss mathematicians Jacques Bernoulli (1654-1705) and his brother Jean Bernoulli (1667-1748), the Leibniz calculus concepts were refined into a convenient scientific tool that became widely used.

It was the combination of this calculus tool developed by Leibniz and the Bernoulli brothers, along with Newton's laws, that gave the world a practical engineering approach to mechanics. Another scientific achievement of Leibniz was his invention in 1672 of a calculating machine for multiplying, dividing and taking square roots.

Calculation of the Acceleration of Gravity

Let us apply Newton's laws to calculate the acceleration of gravity, which is denoted g. In Table 4-1, item (1) gives the general formula for gravitational force between two bodies (a) and (b), which is specified by Newton's law (1), given earlier. Parameter M_a is the mass of body (a), M_b is the mass of body (b), and d_{ab} is the distance between the centers of bodies (a) and (b). Parameter G is the constant of proportionality of the equation, which is called the *gravitational constant*.

Item (1) shows that the gravitational force is proportional to the product of the two masses (M_a and M_b), and is inversely proportional to the square of the distance (d_{ab}) between the two bodies.

In item (2), the general formula of item (1) is applied to obtain the weight (W) of an object. The weight is the gravitational force exerted by the earth on an object at the surface of the earth. Parameter M is the mass of the object, M_e is the mass of the earth, and r_e is the radius of the earth, which is the distance between the center of the earth and the center of the object.

Table 4-1: Derivation of formula for acceleration of gravity

Characteristic	Value
(1) General law of gravitational force	$GM_aM_b/(d_{ab})^2$
(2) Weight (W) of object of mass M on earth	$GMM_e/(r_e)^2$
(3) Acceleration (g) of falling object (W/M)	$GM_e/(r_e)^2$

When this object is allowed to fall, the gravitational force W causes the object to accelerate downward with the acceleration of gravity (g). By Newton's law (2), the gravitational force W is equal to the mass M of the object multiplied by the acceleration of gravity (g). Hence, (g) is equal to the ratio (W/M), as shown in item (3). Dividing item (2) by M gives the formula for the acceleration of gravity shown in item (3).

The value for the acceleration of gravity (g) was known in Newton's time, and the radius of the earth (r_e) had been measured. Hence, Newton was able to calculate from the formula of item (3) the value of the product (GM_e). However, Newton had no direct means of finding the mass of the earth M_e, and so he could not determine the actual value for the gravitational constant G.

In 1798, Henry Cavendish measured the gravitational constant G directly by sensing the gravitational attraction between two lead spheres. When G was known, the mass of the earth M_e could be calculated. Hence it was said that, "Cavendish weighed the earth in his experiment". The Cavendish experiment is described later in this chapter.

The Orbits of Planets

The gravitational pull of the sun on the earth holds the earth in its orbit as it revolves around the sun. In accordance with Newton's fourth law that for every action there is an equal and opposite reaction, the earth exerts the same force on the sun as the sun exerts on the earth. The gravitational pull of the earth on the sun moves the center of the sun slightly as the earth moves in its orbit around the sun.

You can illustrate the motion of the earth in its orbit by swinging a ball on the end of a string. You are the sun and the ball is the earth. The tension in the string represents the gravitational attraction between the sun and the earth. The force that the string applies to the ball is the gravitational force that holds the ball (earth) in a circular trajectory. If you let go of the string, the ball flies away in a straight line (except for the downward drop due to gravity). Thus the gravitational pull from the sun keeps the earth in its orbit, and prevents it from flying off into space.

The gravitational force that the sun applies to the earth causes the earth to accelerate. However this acceleration does not change the absolute value of the earth velocity (which we call *speed*), it merely changes the direction of the earth velocity.

This concept is illustrated in Fig. 4-2. At time (1) the earth velocity is shown by the arrow V_1, which is called a "vector". The direction of the V_1 vector arrow shows the direction of the earth velocity at that instant, and the length of the vector arrow is proportional to the speed of the earth (30 km/sec).

At time (2) the earth velocity is shown by vector arrow V_2. The velocity vectors show the instantaneous values of velocity at points (1) and (2). The lengths of the two velocity vectors are the same, because the absolute value of the earth velocity (the *speed*) is constant. However, the direction of the velocity vector changes. Each velocity vector is tangent to the circular orbit at the instantaneous location of the earth.

In diagram (b), vectors V_1 and V_2 are moved without changing direction, so that they start at the same point. Vector ΔV is the difference between these two vectors, and is constructed by drawing a vector from the tip of vector V_1 to the tip of vector V_2. Vector ΔV is the amount by which the earth velocity changes between points (1) and (2). The Greek letter Δ (delta) is commonly used to denote a difference; in this case ΔV denotes a difference in velocity V.

Acceleration is the change of velocity divided by the time interval over which the velocity change occurs. Dividing the length of the velocity difference ΔV by the time between (1) and (2) gives the average acceleration between points (1) and (2). The earth accelerates between points (1) and (2) even though the speed of the earth stays constant.

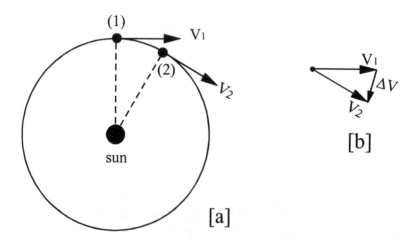

Figure 4-2: Change in the velocity of a planet as it rotates around the sun in a circular orbit

This shows that velocity must be treated as a vector when acceleration effects are calculated. A vector has both direction and amplitude. If the acceleration produces a velocity change in the direction of the velocity vector, the speed changes. (That case was illustrated by the falling object in Fig. 4-1.) If the acceleration produces a velocity change perpendicular to the velocity vector (which is the case in Fig 4-2), the velocity direction changes but the speed does not.

This concept is illustrated in Fig. 4-3. In diagram (a) the velocity change ΔV is in the direction of the velocity vector V_1. Consequently, as shown by the sum velocity V_2, the velocity difference increases the length of the velocity vector and thereby increases the speed.

In diagram (b) the velocity change ΔV is perpendicular to the direction of the velocity vector V_1. Consequently, as shown by the sum velocity V_2, the velocity difference ΔV changes the direction of the velocity vector but not its length, and so does not change the speed. (There is a slight difference of length in diagram (b), but this effect becomes negligible for small velocity changes.)

Newton combined his laws of motion with calculus and showed that they exactly satisfied all three of Kepler's Laws. The accurate empirical laws derived by Kepler were essential elements in the development of Newton's theory. One of the greatest tragedies of scientific history was that Johannes Kepler died in poverty, despite the revolutionary character of his discoveries.

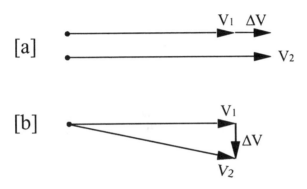

Figure 4-3: Effect of changing a velocity vector: [a] change in direction of velocity vector; [b] change perpendicular to velocity vector

Orbit of the Moon

Newton applied his laws to the orbit of the moon around the earth, and derived a formula that related the following parameters:

(1) Acceleration of gravity g on the earth
(2) Radius of the earth
(3) Velocity of the moon
(4) Distance from center of earth to center of moon

The values for all of these parameters were known to Newton. Newton demonstrated that his formula was satisfied within measurement error and thereby provided strong evidence that his theory was valid. Newton withheld publication of his *Principia* until the latest measurements of distance to the moon and the diameter of the earth agreed with his theory.

Newton's laws have predicted with very high accuracy the motions of planets, moons, comets, and other bodies in our solar system. It was not until 1916 that Einstein proved with his General theory of Relativity that Newton's gravitational theory is not absolutely accurate. Einstein showed that relativistic effects produce a tiny error in the orbit of the planet Mercury that cannot be calculated from Newton's theory

Why Are Astronauts Weightless?

Why do astronauts in space experience weightlessness? Although this phenomenon is well known today, many people do not understand why it occurs.

Table 4-1 showed that the acceleration of gravity on the surface of the earth is $GM_e/(r_e)^2$. At the location of a space vehicle in orbit around the earth, the parameter r_e in this formula is replaced by the radial distance r measured from the center of the earth to that point. Thus, the acceleration of gravity g is equal to GM_e/r^2, where r is equal to $(r_e + h)$. Parameter r_e is the radius of the earth (6378 km), and h is the altitude of the space vehicle.

The International Space Station orbits the earth at an altitude of 418 km (260 miles). At this altitude the acceleration of gravity is 88 percent of the value at the surface of the earth. Hence the gravitational force pulling an astronaut in the space vehicle toward the center of the earth is 88 percent of the weight that the astronaut experiences on the ground. Nevertheless, the astronaut is weightless in space. Why?

The reason for the weightlessness is that the space vehicle and the astronaut are in a free-fall condition. They are continually accelerating toward the center of the earth with an acceleration of gravity g corresponding to that altitude. However, the vehicle is traveling so fast that the altitude of the space vehicle stays constant. The acceleration merely changes the direction of the velocity vector. An object at the 418 km altitude of the International Space Station stays at constant altitude if its velocity V is 7.67 km/sec (4.77 mile/sec).

Thus the space vehicle and the astronaut are continually falling toward the center of the earth, even though their altitude does not change. In this free-fall condition, the astronaut is weightless.

A sky-diving parachutist experiences weightlessness for a few seconds after jumping out of an airplane, until the velocity is sufficient for air resistance to offset the force of gravity. The wind force from the air is proportional to the square of velocity. It matches the sky diver weight at a speed of about 120 mph when the sky diver's body is oriented horizontally, or 180 mph when the body is oriented vertically. During the first few seconds the wind force is very small, and so the falling sky diver is essentially weightless.

A simple way to experience temporary weightlessness is to jump from a high platform into water, or take a carnival parachute drop.

How Cavendish Weighed the Earth

Although we realize that we are continually being pulled downward because of the gravitational pull of the earth, we are not aware that we are being pulled in other directions by the gravitational pull of other bodies on earth, because the gravitational tug from another body is extremely tiny. Nevertheless, it is still present, and Cavendish used this principle to measure the gravitational constant G in Newton's law of gravitation.

In 1798 Henry Cavendish (1731-1810) measured the gravitational constant G by sensing the gravitational attraction between lead spheres. Henry Cavendish was a brilliant and wealthy scientist, but was reclusive and eccentric. He is best known for his chemistry research, which included the discovery that water is a compound of hydrogen and oxygen.

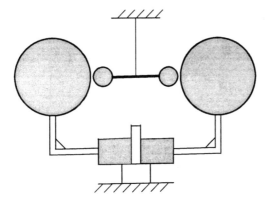

Figure 4-4: The Cavendish experiment to measure Newton's gravitational constant G

This gravitational experiment by Henry Cavendish is shown in Fig. 4-4. A horizontal dumbbell, with lead balls at each end, was suspended from a thin vertical wire. A pair of much larger lead spheres was mounted on a rotatable structure, which moved the spheres just outside the lead balls. As the large spheres were moved past the balls of the dumbbell, the gravitational attraction between the dumbbell balls and the large spheres made the dumbbell rotate, and caused the wire to twist slightly. The amount of twist was sensed optically to give a measure of the gravitational attraction.

The weight of a dumbbell ball was about 10 million times greater than the tiny gravitational force between the ball and the large sphere. Nevertheless the experiment was sufficiently sensitive to detect this gravitational force. From this experiment, Cavendish measured the gravitational constant G within an error of 1.4 percent.

Table 4-1 showed that the acceleration of gravity g is equal to GM_e/r_e^2, where M_e and r_e are the mass and radius of the earth. Since the values of g and r_e were known, this measurement of the gravitational constant G allowed Cavendish to calculate the mass of the earth M_e. Consequently it was said that, *"Henry Cavendish weighed the earth in his experiment"*. He found that the average density of the earth is 5.5 times that of water.

Chapter 5

The Nature of Light

The theory of Relativity evolved from a study of the speed of light. To introduce this issue, this chapter explains the physical principles of light propagation.

What Is a Light Wave?

We know that light and sound are waves. We can help to understand these waves by examining the propagation of a wave on water.

Mechanical Waves

We can envision a wave by dropping a stone into a smooth pond. The water waves propagate outward from the point where the stone hits the water. An individual water particle oscillates back-and-forth and up-and-down, following an elliptical path. It is the energy of the wave that propagates across the pond, not the water itself. This effect can be observed by noticing that an object floating on the surface moves very little as the wave passes it.

If we drop two stones into the pond at once, two sets of waves are produced, and these two sets of waves interfere with one another. Similar wave interference occurs with sound and light waves.

Like a wave on water, sound is a mechanical vibration. One can feel the vibration in a musical instrument when a tone is produced. The instrument vibrates the air, producing sound that travels through the air to the ear that hears it. Sound is a compression wave, in which the air particles vibrate back and forth in the direction of propagation of the sound.

Electromagnetic Waves

A wave on water and a sound wave are mechanical waves, which propagate by vibrating a medium. Although a light wave is similar in certain respects to these mechanical waves, it is fundamentally very different. A light wave is a packet of oscillating electric and magnetic fields, and so is called an electromagnetic wave.

A light wave is the same as a radio wave except that it oscillates at a much higher frequency. Standard AM (amplitude modulation) radio operates at a frequency of about 1 megahertz (MHz). One hertz (Hz) means one cycle per second, and so 1 MHz means 1 million cycles per second. Television operates at a frequency of about 500 MHz; light has a frequency of about 500 million MHz, and an X-ray has a frequency of about 3 trillion MHz.

To help understand an electromagnetic wave, let us consider some examples of electric and magnetic fields. We observe the effect of an electric field in a thunderstorm. The severe winds of the storm remove electrons from the ground and deposit them in the clouds. The clouds become charged negatively with respect to the ground, and a large electric field builds up between the clouds and the ground. Eventually this electric field gets so strong that an electric spark (called lightning) jumps between a cloud and the ground. Electrons flow from cloud to the ground, and temporarily eliminate the electric field.

Our earth has a magnetic field, which we can sense with a magnetic compass. A magnetic compass is a magnet that is free to turn and points in the direction of the earth's magnetic field. Our earth acts like a huge magnet, which has north and south poles that are close to the poles about which the earth rotates. Consequently a magnetic compass points approximately in the direction of true north.

Although electric and magnetic fields are quite different, they are closely related, as shown by the following:

(1) A changing magnetic field produces an electric field
(2) A changing electric field produces a magnetic field

Property (1) can be observed in an automobile generator, which delivers electric current to charge the battery. The rotating part of the generator, called the rotor, acts as a magnet. As the rotor spins, its magnetic field moves through a coil of wire on the fixed part of the generator (the stator). This action produces an electric field that generates an electric current to charge the battery. *Hence, by property*

(1), a changing magnetic field produces an electric field.

Property (2) is observed in an electromagnet. Flowing electrons produce an electric current, which is a moving electric field. Feeding the current through a coil of wire forms an electromagnet, which generates a magnetic field. The current in the coil is a moving (or changing) electric field, and this changing electric field produces a magnetic field. *Hence, by property (2), a changing electric field produces a magnetic field.*

Principle of an Electromagnetic Wave

An electromagnetic wave consists of oscillating (or changing) electric and magnetic fields. By property (1), the oscillating (or changing) magnetic field produces an oscillating electric field, and, by property (2), the oscillating (or changing) electric field produces an oscillating magnetic field.

An electromagnetic wave is illustrated in Fig. 5-1. Electric and magnetic fields oscillate at right angles to one another. The oscillating electric field produces the oscillating magnetic field, and the oscillating magnetic field produces the oscillating electric field. These two oscillating fields form a wave that propagates at the speed of light.

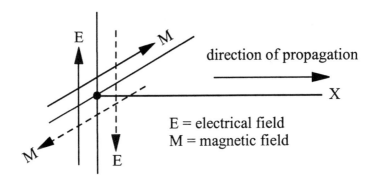

Figure 5-1: Oscillating electric and magnetic fields that form an electromagnetic wave.

The wave propagates in a direction that is perpendicular to the plane within which the electric and magnetic fields oscillate. The electric field E vibrates up and down in the vertical direction. The magnetic field M vibrates back and forth in the horizontal direction. The solid E arrow shows the direction of the electric field during one half cycle, and the

68 The Mystery of Creation

dashed E arrow shows the direction in the next half cycle. Similarly, the solid and dashed M arrows show the directions of the magnetic field in alternate half cycles.

The oscillating electric and magnetic fields are out of phase with one another, so that when one field is a maximum the other is zero. The oscillating electric and magnetic fields support one another, to form a packet of energy that propagates in the horizontal X-direction. The plane in which the electric and magnetic fields oscillate is perpendicular to the X-direction in which the wave propagates.

In 1873, James Clerk Maxwell (1831-1879) derived a set of mathematical equations that combined electrical and magnetic effects. From these equations Maxwell predicted the electromagnetic wave. From measurements of electrical and magnetic parameters, Maxwell calculated the propagation speed of this electromagnetic wave, and found that it was equal to the speed of light. Consequently Maxwell concluded that light must be an electromagnetic wave of very high frequency.

Maxwell's prediction of the electromagnetic wave in 1873 led to the discovery in 1888 of the radio wave by Heinrich Hertz (1857-1894). Seven years later, in 1895, Guglielmo Marconi (1874-1937) used a radio wave of much lower frequency in the first practical radio communication system, when he was only 21. (*Guglielmo* is Italian for *William* and is pronounced *"Gulliamo"*.) In 1903, Marconi achieved telegraphic radio communication from England to Cape Cod, Massachusetts.

Light is a packet of electric and magnetic fields that travels at the speed of light (300,000 km/sec). Since the electric and magnetic fields oscillate as they propagate, light acts like a wave. However, light is not like a sound wave or a wave on water, which propagate by vibrating a medium. There is no medium involved in light propagation. Light can travel through a vacuum, but sound cannot because there is no medium in the vacuum to vibrate.

Measuring the Speed of Light

The speed of sound. Relativity theory evolved from an enigma associated with measuring the speed of light. To help understand what we mean by the speed of light, let us first consider the simpler problem of measuring the speed of sound.

Sound travels at a velocity of 340 meters per second relative to the air. Assume that we measure the velocity of sound with equipment mounted on the ground. If there were no wind, the equipment would measure a sound velocity of 340 meters per second. If the wind were blowing at 20 meter/sec in the direction of the sound wave, the equipment would measure a sound velocity of (340 + 20) or 360 meter/sec. If the wind were blowing at 20 meter/sec opposite to the sound wave, the equipment would measure a sound velocity of (340 - 20) or 320 meter/sec.

The speed of light. Now let us consider light propagation. The wave property of light was demonstrated in the early 1800's by Thomas Young (1773-1829) and Augustin Fresnel (1788-1827). It was originally believed that light travels by vibrating a mysterious medium called the aether (or ether), just as sound travels by vibrating the air. Then in 1873 Maxwell's theory proved that light does not vibrate the aether. *Nevertheless, it was still widely believed that the aether provided an absolute reference, relative to which the light wave propagated.*

Light travels about one million times faster than sound. The velocity of light is 300 million meters per second, or 300 meters per microsecond. One microsecond is one millionth of a second.

Since sound travels with respect to the air, the velocity of the sound wave is measured relative to the air. What happens with a light wave? Suppose that the light emitter and light receiver are moving relative to one another at a velocity that is appreciable relative to the speed of light. To what reference is the speed of light measured? Is the light speed measured relative to the emitter; or is it measured relative to the receiver; or (like sound) is it measured relative to the aether medium through which the light is propagating?

Figure 5-2 helps to clarify this issue. Let us consider the light emitted from a star that is traveling toward earth at an absolute velocity of 300 km/sec. Assume that the earth is traveling toward the star at an absolute velocity of 200 km/sec, so that the relative velocity between the earth (receiver) and the star (emitter) is 500 km/sec.

Figure 5-2 shows the velocities that one *might* calculate for this application. Let us assume that light travels at the speed of light (300,000 km/sec) relative to the emitter. Since the emitter (star) is moving at 300 km/sec, the absolute velocity of the light wave should be the sum of the two velocities, which is 300,300 km/sec.

Since the earth is traveling toward the star at an absolute velocity of 200 km/sec, one might expect the relative velocity between the light wave and the earth to be (300,300 + 200) or 300,500 km/sec. This reasoning indicates that the relative speed of the light received from a star should be equal to the nominal speed of light (300,000 km/sec) plus the relative velocity between the star and the earth (500 km/sec for this example).

But this conclusion is not true! The velocity of the light that we receive from a star is always exactly the same, 300,000 km/sec. We can tell from the Doppler wavelength shift in a star's spectrum that the relative velocity between a star and earth varies greatly from star to star. Nevertheless the speed of the light that we receive from a star is unaffected by the relative velocity of the star. How do we explain this confusing result?

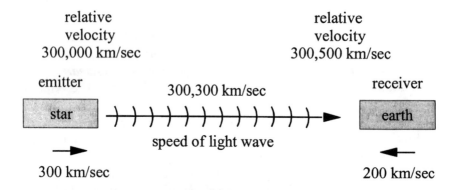

Figure 5-2: Assumed speed relations associated with the propagation of a light wave

In the late 1800's, experiments found contradictions in measuring the speed of light. To explain these, it was widely believed that light must move at a constant velocity relative to the aether, just as sound moves at a constant velocity relative to the air.

The Michelson-Morley Experiment

In 1887, Albert Michelson (1852-1931) and Edward Morley (1838-1923) implemented the famous Michelson-Morley experiment, which was performed to measure the velocity of the aether. The earth revolves around the sun at a velocity of 30 km/sec, which is 0.01 percent of the speed of light (300,000 km/sec). During the year the earth moves at ±0.01 percent of the speed of light in different directions, and so should move at ±0.01 percent of the speed of light relative to the aether.

Michelson and Morley built a very accurate instrument, using wave interference effects, that could measure differences in the speed of light in perpendicular directions to an accuracy much better than 0.01 percent of the speed of light. The instrument was slowly rotated so that a given path was sometimes in the direction of the earth's motion, and sometimes perpendicular to that motion.

The instrument was applied many times during the year, but there was no detectable difference of the speed of light in the two perpendicular paths, regardless of the direction of the earth's motion. The results of the Michelson-Morley experiment were completely negative.

Many hypotheses were proposed to explain the negative results of the Michelson-Morley experiment. It was suggested that in the vicinity of the earth the aether moves with the earth, just like the earth's atmosphere. However, if this were true, the aether in various parts of our solar system would have different velocities, and so the speed of light would be different. If this occurred, the variation of the speed of light across our solar system would be obvious.

The Contraction Hypothesis

Finally, Professors George Francis FitzGerald (1951-1901) and Hendrik Lorentz (1853-1928) independently proposed the hypothesis that the length of an object contracts when the object moves relative to the aether. A dimension in the direction of the relative velocity should contract by the following factor K:

$$K = \sqrt{[1 - (V/c)^2]}$$

where V is the velocity of the object relative to the aether, and c is the speed of light. As the Michelson-Morley instrument rotates, the two arms should change in length according to this formula, and this postulate can explain the negative results of the experiment. [7]

This contraction effect can be represented in a geometric manner as shown in Fig. 5-3. The hypotenuse AB of the right triangle has a length of unity. The length of the side BC is equal to the velocity ratio V/c. The third side AC is equal to the contraction factor K. Dimensions are assumed to contract by the factor K, which is the ratio of side AC divided by the hypotenuse AB.

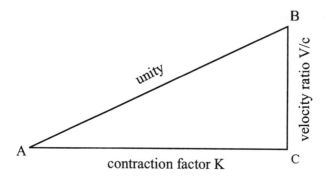

Figure 5-3: Right triangle showing in a geometric manner the relation between the V/c velocity ratio and the Lorentz contraction factor K

The Lorentz Transformation

Other experiments were performed to measure motion relative to the aether, but all of them had negative results. Lorentz found that he could explain all of these experiments by generalizing his contraction hypothesis. He developed a set of equations postulating that time as well as distance measurements are altered when a body moves relative to the aether. These Lorentz equations were published in 1904, and later became known as the ***Lorentz transformation equations***. [7]

Lorentz proved that the electromagnetic field equations developed by Maxwell would be *"invariant"* if modified by his equations. [7] This result indicated to Lorentz that it was probably impossible to measure the velocity of a body relative to the aether if the Lorentz transformation equations are valid.

But if this is true, how can the aether have any meaning? If it is impossible to detect the presence of the mysterious aether, why should we believe that the aether actually exists?

The Einstein Relativity Principle

Into this confusion stepped the brilliant young physicist, Albert Einstein (1879-1955). In 1905 Einstein [8] published his paper on Relativity, the year after the Lorentz paper. Einstein approached the enigma of the speed of light by applying fundamental reasoning. He concluded that *absolute velocity* has no meaning; there is only *relative velocity.* ***We can specify the relative velocity between two bodies, but*** *not the absolute velocity of either one.*

Based on this principle, Einstein concluded that the aether cannot exist. Since absolute velocity has no meaning, there cannot be an aether medium that establishes an absolute reference for specifying velocity.

How do we specify the speed of light? Let us refer back to Fig. 5-2. The light travels from the emitter (star) to the receiver (earth). Since the speed of light can only be specified in a relative sense, we can consider (1) that the speed of light is the relative velocity between the emitter and the light, and (2) that the speed of light is the relative velocity between the receiver and the light.

Since absolute velocity has no meaning, the velocity of an observer located at the emitter is no more (nor no less) preferred than the velocity of an observer located at the receiver. Consequently, observers located at the emitter and receiver must measure exactly the same relative velocity for the speed of light.

Thus, Einstein established the following principle: ***Two observers moving at different velocities must measure exactly the same value for the speed of light, regardless of the relative velocity between them.***

Einstein then asked, "What conditions must be satisfied in order for this principle to hold?" Einstein concluded that if there is a relative velocity between two observers, their measurements of distance and time must be different. [8]

This means, for example, that an object does not have a definite physical length. The length of an object depends on the velocity of the observer that is measuring it. Dimensions are relative. Time intervals are relative. ***Reality is Relative.***

This is the essence of the Einstein theory of Relativity. In Chapter 6 the Einstein Relativity theory is explained in detail.

Reaction to the Einstein Relativity Theory

Einstein had found the key to the enigma concerning the speed of light that was confusing the scientific world. Einstein used the same *Lorentz transformation equations* that had been derived a year earlier by Lorentz, but interpreted them in a radically different manner. With this new interpretation, Einstein made profound physical predictions that were not achieved by Lorentz.

In 1905, when Albert Einstein published his paper on Relativity, he was an unknown physicist working in the Swiss Patent Office. In that same year Einstein published three other landmark papers. [23] (pp. 120-121) One paper dealt with the photoelectric effect, which proved that light is quantized into small units, called photons. Another paper determined the true size of molecules from their diffusion in a diluted liquid solution. A third paper analyzed the statistics of the motions of gas particles, which established Einstein as the founder of statistical mechanics. Einstein's Relativity paper was the fourth of these landmark papers on physics published in 1905.

Einstein received the Nobel Prize for his paper on the photoelectric effect in 1922. He did not receive the Nobel Prize for his much more important research on Relativity, because it was too controversial at the time. Einstein's paper on the photoelectric effect proved that light not only acts like a wave; it also acts like a stream of particles.

Chapter 6

Einstein Special Theory of Relativity

This chapter explains the basic theory of Relativity that Einstein presented in 1905. When Einstein presented his General theory of Relativity in 1916, this basic theory became known as the Special theory of Relativity.

Chapter 5 explained the fundamental principle underlying Einstein's theory of Relativity. Einstein concluded that absolute velocity does not exist; only relative velocity has meaning. Consequently light must move at the same relative velocity with respect to all observers. Two observers moving at constant velocity must measure exactly the same speed of light, regardless of the relative velocity between them. To satisfy this requirement, Einstein showed that the measurements of distance and time must be different for the two observers.

We start our explanation of Relativity by examining the process of measuring the speed of light.

Measuring the Speed of Light

Light travels 300 million meters per second. This can be expressed as 300 meters per microsecond, where one microsecond is one-millionth of a second. The speed of light is measured by finding the time for light to travel the length of a measuring rod.

As shown in Fig 6-1, let us consider a 3-meter measuring rod. Since light takes one microsecond to travel 300 meters, it travels 30 meters in 0.1 microsecond, and it travels 3 meters in 0.01 microsecond.

Therefore light takes 0.01 microsecond to travel the length of a measuring rod that is 3 meters long. It is convenient to express this time in terms of the nanosecond, which is one-thousandth of a microsecond or one billionth of a second. Light takes 10 nanoseconds (or 0.010 microsecond) to travel the length of a 3-meter measuring rod.

Because light travels so fast, special instrumentation is needed to measure the time for light to travel the length of a measuring rod. This can be achieved as shown in Fig. 6-1. Clocks 1 and 2 are placed at the leading and trailing ends of the measuring rod. The two clocks run at exactly the same rate and are accurately synchronized. Synchronization of the clocks could be achieved by locating the clocks at the same point, setting their readings equal, and then moving them to the ends of the measuring rod

When the light reaches the leading edge of the measuring rod, clock 1 is read, and when the light reaches the trailing edge, clock 2 is read. The clock readings are subtracted to obtain the time for the light pulse to travel the length of the measuring rod. The difference between the clock readings is 10 nanoseconds.

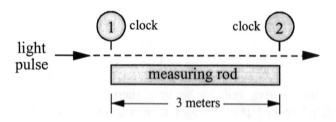

Figure 6-1: Measuring the speed of light

The Einstein Theory of Relativity

To illustrate the Einstein theory, let us apply it to a fictitious space travel episode. This example yields relativistic effects that are large enough to be conveniently evaluated.

As shown in Fig 6-2, a space ship is returning from an interstellar voyage, and is now traveling toward earth at 60 percent of the speed of light. Since the speed of light is 300 meters per microsecond, the space ship velocity is 180 meters per microsecond.

A light pulse is transmitted from earth, which leaves earth at a speed of 300 meters per microsecond. The earth observer A and the space ship observer B measure the speed of this light pulse. Observers A and B measure exactly the same value for the speed of light. Both observers have identical equipment, and both find that it takes exactly 10 nanoseconds for the light to travel the length of a 3-meter measuring rod.

6. Einstein Special Theory of Relativity

In terms of our fanciful example, let us see how Einstein explained the constancy of the speed of light in his 1905 paper on Relativity.

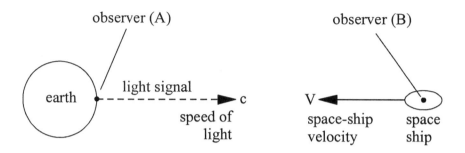

Figure 6-2: Measuring the speed of a light signal transmitted from earth to a space ship, which is returning at a velocity V equal to 60 percent of the speed of light c.

The Principles of Relativity

Relative to the earth observer *A*, a measuring rod on the space ship appears to contract by the contraction factor K discussed in Chapter 5. Since the relative velocity V between the earth and the space ship is 60 percent of the speed of light c, the ratio V/c is 0.6 and the contraction factor K is

$$K = \sqrt{[1 - (V/c)^2]} = \sqrt{[1 - (0.6)^2]} = \sqrt{[0.64]} = 0.8$$

In accordance with Fig 5-3 of Chapter 5, this contraction factor K can be expressed geometrically as shown in Fig 6-3.

The hypotenuse AB of the right triangle in Fig. 6-3 is unity. Side BC is equal to the velocity ratio V/c, which is 0.6. The third side AC is equal to the contraction factor K, which is 0.8. This example was chosen to achieve a right triangle with the well-known ratios 6:8:10 (or 3:4:5).

To the earth observer *A*, the measuring rod on the space ship appears to contract to 80 percent of its normal length when oriented in the direction of relative velocity. To the earth observer *A*, the 3-meter measuring rod on the space ship appears to be 2.4 meters long, which is 80 percent of 3 meters. A dimension perpendicular to the velocity direction is not changed.

The clocks on the space ship appear to run slow when observed from earth. To the earth observer A, the clocks on the space ship appear to run at 80 percent of the rate observed by B on the space ship. A clock rate is compressed by the same contraction factor K as a spatial dimension.

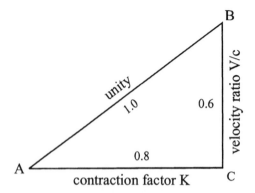

Figure 6-3: Right triangle showing in a geometric manner the relation between the V/c velocity ratio and the Special Relativity contraction factor K

The third and final relativistic effect is that the two clocks on the space ship, which observer B considers to be synchronized, are not synchronized relative to the observer A on earth. The synchronization error in these clocks seen by observer A is

Synchronization error = (V/c)Δt = 0.6 Δt

The quantity Δt is the time for light to travel between the two clocks, as seen by observer B. Observer B finds that it takes 10 nanoseconds for light to travel the length of his measuring rod, and so Δt is 10 nanoseconds. Hence observer A sees a clock synchronization error that is 60 percent of 10 nanoseconds, which is 6 nanoseconds. This synchronization error is measured in terms of the space-ship clocks.

Observer B on the space ship considers his two clocks to be exactly synchronized, but observer A on earth considers the two clocks to be out of synchronization by 6 nanoseconds, as measured on the space-ship clocks.

Explanation of Constancy of the Speed of Light

Let us apply the above principles to explain why observers A and B measure exactly the same value for the speed of light.

6. Einstein Special Theory of Relativity

To observer A on earth, the light pulse emitted from earth should travel relative to the space ship at a velocity equal to (c + V), which is (300 + 180), or 480 meters per microsecond. To observer A, the spaceship measuring rod appears to be 2.4 meters long. Hence the time for the light pulse to travel the length of the measuring rod is equal to the rod length (2.4 meters) divided by the speed of light relative to the space ship (480 meters per microsecond).

Relative to observer A, the light pulse takes 2.4/480 microsecond to travel the length of this measuring rod. This is 1/200 microsecond, or 5 nanoseconds. Therefore to observer A the light pulse appears to take 5 nanoseconds to travel the length of the measuring rod on the space ship.

However, the clocks on the space ship appear to run at 80 percent of the normal rate. Consequently, observer B should interpret this 5-nanosecond time interval to be 4 nanoseconds. From the point of view of observer A, it should take 4 nanoseconds for the light to travel the length of the measuring rod as measured on the space-ship clocks.

To the earth observer A, the two clocks used by B are out of synch by 6 nanoseconds. This 6-nanosecond synchronization error should be added to the 4-nanosecond time interval to obtain the measured time for light to travel the length of the measuring rod. Therefore from the point of view of earth observer A, observer B on the space ship mistakenly thinks that it takes (6 + 4) nanoseconds, or 10 nanoseconds, for light to travel the length of his measuring rod.

In this manner we can explain why both observers find that it takes exactly 10 nanoseconds for light to travel the length of the measuring rod. Consequently both observers measure the same speed of light.

Symmetry of Relation between Observers A and B.

Absolute velocity does not exist; there is only relative velocity. The earth observer A can assume that the earth is stationary and the space ship is moving at 60 percent of the speed of light. Likewise, the space ship observer B can assume that the space ship is stationary, and the earth is moving at 60 percent of the speed of light.

Let us assume that the space ship sends a light pulse to earth. The preceding discussion can then be applied to the space ship rather than to the earth, by considering the space ship observer to be A and the earth observer to be B. The measurements of the speed of light are symmetric.

Replacing the Observer by a Set of Coordinates

Instead of considering measurements made by an observer, we can consider measurements made relative to coordinates at the location of the observer. The principles of Relativity show how the measurements made with respect to different coordinates are related to one another.

Relativity of observation. In comparing the apparent effects experienced by the earth and space-ship observers, we should recognize that these apparent effects are real. *Reality is Relative.* There are no absolutes. The measuring rod on the space ship is 3 meters long to the space ship observer, and it is 2.4 meters long to the earth observer. Both statements are correct. The measuring rod does not have an absolute length.

Proper coordinates. On the other hand, we can still consider the rod length measured on the space ship to be special, and we call this the *proper length* of the rod. Coordinates that move with the space ship are called *proper coordinates* for the space ship. The length of the space ship measuring rod relative to these proper coordinates is called the *proper length* of the rod. The *proper length* of the space-ship measuring rod is 3 meters.

The measuring rod on earth is 3 meters long relative to the earth observer, and 2.4 meters long relative to the space ship observer. The proper length of this measuring rod is 3 meters, which is the length that is measured from proper coordinates moving with the earth.

Four Dimensionality of Space and Time

The fact that two clocks that are synchronized relative to one observer are not synchronized relative to another observer tells us that there is no such thing as absolute time. Events that are simultaneous to one observer are generally not simultaneous to another observer moving at a different velocity.

This indicates that measurements in time and space cannot be considered separately. In Relativity, time and spatial measurements are combined together into a *four-dimensional space-time specification.* This does not mean that we should regard time to be a mysterious fourth dimension that is equivalent to a spatial dimension. To any observer, space and time measurements are radically different concepts.

The four dimensionality of space-time means that time and space must be considered together to obtain a precise specification. For example, a time interval between two events that is experienced by one observer can appear to be a distance interval to another observer moving at a different velocity.

Variation of Mass of an Object

Let us consider what happens when an object is accelerated until its velocity gets close to the speed of light. This effect can be achieved experimentally by accelerating an electron in an electric field, as shown by the plots in Fig. 6-4. The electric field applies a constant force to the electron. The solid curve shows the resultant velocity of the electron relative to the speed of light.

Initially the electron velocity increases at a constant rate. However, when the electron velocity V nears the speed of light (when V/c is close to unity) the slope of the velocity plot decreases, and the velocity approaches the speed of light gradually.

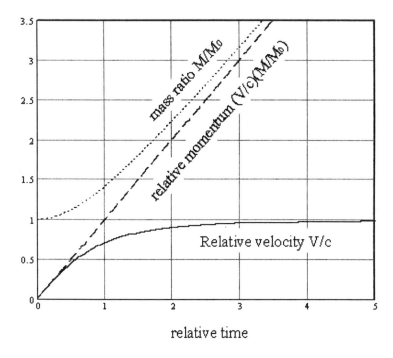

Figure 6-4: Variation with time of relative velocity, relative momentum, and relative mass of an electron accelerated by a constant force to nearly the speed of light

The momentum of a body is the product of its mass M times its velocity V. The force applied to a body is equal to the rate of change of the body's momentum. Since a constant force is applied to the electron, its momentum must increase at a constant rate. The dashed line shows how the momentum of the electron increases with time.

In order for the momentum (MV) of the electron to increase at a constant rate (even though the velocity V approaches the speed of light gradually), the mass M of the electron must increase. The dotted curve shows how the mass of the electron increases as the electron velocity approaches the speed of light.

The dotted curve is the ratio M/M_0, which is the ratio of the electron mass M divided by its *rest mass* M_0. Rest mass is the electron mass at zero velocity. The mass of the electron traveling at the velocity V is

$$M = M_0/\sqrt{[1 - (V/c)^2]}$$

According to this formula, the mass M would be infinite if the velocity V reached the speed of light c. ***Therefore the formula shows that nothing with mass can travel at the speed of light.***

Proper Coordinates for the Electron

Let us consider *proper coordinates* that move with the electron. Relative to these proper coordinates, the mass of the electron is its rest mass, which does not vary. If we could carry a clock in these proper coordinates, the clock would run slower as the electron velocity nears the speed of light. This effect has been observed experimentally by accelerating radioactive particles. The rate of radioactive decay of these particles decreases when the velocity of the particles gets close to the speed of light.

When the electron is close to the speed of light, its proper clock would run very slowly. If the electron could travel at the speed of light, its proper clock would stop. The electron would then travel an infinite distance (as measured externally) in zero *proper time* (as measured on a clock traveling with the electron).

Converting Matter into Energy

The increase in mass of an object as it approaches the speed of light indicated to Einstein that adding energy to an object increases its mass. This in turn showed that mass M can be converted into energy E in accordance with the following famous Einstein formula:

$$E = Mc^2$$

This formula explained the source of the energy radiated by the sun, and eventually led to the atomic and hydrogen bombs. The equation shows that one gram of mass M is equivalent to 25 million kilowatt-hours of energy E. This means that 25 million kilowatt hours of energy are released for every gram of matter that is converted into energy.

The sun achieves its energy by fusing four hydrogen atoms to form one helium atom. The atomic weight of helium is 3.9716 times that of hydrogen, and so the ratio of helium mass to hydrogen mass is 3.9716/4, which is 0.99290. The helium atom has 99.290 percent of the mass of the four hydrogen atoms that form it. The remaining 0.710 percent of the hydrogen mass is converted into energy. Taking 0.71 percent of 25 million kilowatt-hours gives 177,500 kilowatt-hours. This shows that the conversion of one gram of hydrogen into helium releases 177,500 kilowatt-hours of energy. One gram is 1/3 of the weight of a United States penny (one cent piece).

It is remarkable that the abstract principles of Special Relativity allowed Einstein to explain the source of the enormous energy radiated by our sun. *The fact that matter can be converted into energy, in accordance with this Einstein formula, demonstrates that Einstein's Relativity principles are correct. It proves that Reality is Relative.*

Chapter 7

Einstein General Theory of Relativity

Generalizing the Relativity Principle

Einstein derived several profound conclusions from his basic Relativity concept, but he soon found that it had a serious weakness. The theory applies exactly only when the velocities of the two observers are constant. When acceleration occurs, which means that the velocity is changing, Einstein found that the speed of light is not exactly constant. Since constancy of the speed of light is the unifying principle for his basic theory of Relativity, Einstein needed a new principle to generalize his Relativity concept.

Equivalence between Acceleration and Gravity

To calculate the relativistic effects of gravity, Einstein postulated that acceleration and gravity are equivalent. The gravitational pull of the earth exerts a force on our body, which we call weight. In an airplane during take-off, our body is forced back against the seat because of the forward acceleration of the airplane. Einstein postulated that the relativistic effects of gravitation and acceleration forces are equivalent. To generalize his Relativity theory, he applied the principles of Relativity under conditions of acceleration, and related these results to a gravitational field. Let us see how he did this.

In Fig. 7-1, Einstein considered two identical elevators. One in diagram (b) rests on the ground, and the other in (a) is located in space, where gravitational force is negligible. Einstein assumed that the elevator in space is being pulled upward with the same acceleration of gravity g that is experienced on earth. Today we assume more realistically that a rocket motor under elevator (a) is pushing the elevator upward with the acceleration g.

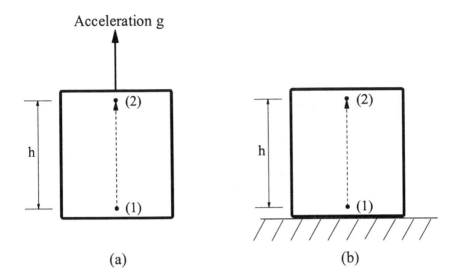

Figure 7-1: Elevator (a) is accelerating in free space; elevator (b) is fixed on earth. Light travels from point (1) to point (2).

The meaning of the acceleration of gravity g was explained in Chapter 4. When an object is allowed to fall on earth, it drops with a constant acceleration of gravity g of about 10 meter/sec per sec. The velocity of a falling object is proportional to time, provided that air resistance is negligible. When an object is not allowed to fall, it exerts a force on the floor, which is called weight. The weight W of an object is equal to its mass M multiplied by the acceleration of gravity g.

Since the elevator in diagram (a) is pushed upward by a rocket with the same acceleration of gravity g experienced on earth, an object within elevator (a) would exert the same weight force on the floor that it would exert if it were located within the fixed elevator (b) located on earth. The conditions inside elevators (a) and (b) are the same. If the elevators are closed, a scientist could not tell from experiments performed within the two elevators whether he is in elevator (a) or elevator (b). This is the *Principle of Equivalence* proposed by Einstein, which allowed him to relate the effects of acceleration and gravity.

Redshift Produced by Gravity

Suppose that a light pulse is emitted from the floor of elevator (a) at point (1) and is received at the ceiling at point (2). During the propagation time of the light, the velocity of the elevator increases, because the elevator is accelerating. Therefore the upward velocity of the receiver at point (2) is greater than that of the emitter at point (1), as far as the light pulse is concerned. Point (2) appears to be moving away from point (1), and so the light is redshifted as it moves from (1) to (2). The wavelength of the light received at (2) is greater than the wavelength emitted at (1).

Table 7-1 shows the steps for calculating this change of wavelength. The calculations of this table are approximate. Item (a) shows the propagation time for light to travel from (1) to (2). The propagation time, which is denoted Δt, is approximately equal to the height h of point (2) above point (1), divided by the speed of light c.

Table 7-1: *Calculation of approximate redshift caused by an accelerating elevator or by a fixed elevator in a gravitational field.*

(a) Propagation time of light (Δt)	h/c
(b) Velocity change during light propagation (ΔV)	$g\Delta t = gh/c$
(c) Velocity ratio ($\Delta V/c$)	gh/c^2
(d) Doppler redshift of light ($\Delta\lambda/\lambda$)	$\Delta V/c = gh/c^2$

During the light propagation time Δt, the velocity at point (2) increases, because the elevator is accelerating. The velocity at point (2), which is denoted V, increases by an amount (ΔV) equal to the elevator acceleration g multiplied by the time interval Δt. As shown in item (b) the velocity increases by an amount equal to $g\Delta t$. Since the Δt time interval is equal to h/c, the velocity increase ΔV is equal to gh/c.

Because of this velocity increase (ΔV), point (2) appears to be moving away from (1), and so the light experiences a Doppler redshift as it travels from (1) to (2). Doppler redshift is the wavelength increase $\Delta\lambda$ divided by the normal wavelength λ, and is approximately equal to

$$\Delta\lambda/\lambda = \Delta V/c \quad \text{(approximate Doppler redshift)}$$

Dividing ΔV in item (b) by the speed of light c, gives the ratio $\Delta V/c$ shown in item (c), which is equal to (gh/c^2). By the above equation, this is approximately equal to the $\Delta\lambda/\lambda$ redshift, as shown in item (d). Thus, the $\Delta\lambda/\lambda$ Doppler redshift is approximately equal to (gh/c^2).

Since the conditions in elevators (a) and (b) are equivalent, a light pulse emitted at point (1) in elevator (b) experiences a redshift when it is received at point (2). The gravitational field in elevator (b) affects the wavelength of light in exactly the same manner as the acceleration applied to elevator (a). Therefore, in accordance with item (d), the gravitational field in elevator (b) produces a $\Delta\lambda/\lambda$ redshift approximately equal to (gh/c^2) as the light propagates from (1) to (2).

Now assume that a light pulse is emitted at the ceiling (point 2) and is observed at the floor (point 1). In this case, point (1) appears to be moving toward (2). A Doppler wavelength shift is observed, but the spectrum is now shifted toward the blue end of the spectrum (toward shorter wavelength). Hence when light travels from the ceiling to the floor, the spectrum experiences a $\Delta\lambda/\lambda$ blueshift approximately equal to (gh/c^2).

Effect of Gravity on Clock Rate

Place a clock at point (1) in elevator (a) of Fig. 7-1, and time the emitted light pulses with the ticking of the clock, so that one pulse is emitted every nanosecond. The spacing between light pulses received at (2) will be greater than one nanosecond. To understand this, imagine that you are moving away from a train of light pulses emitted by a fixed source. As you move, the distance between you and the light source increases, and so the number of light pulses along the transmission path increases. Consequently, the rate at which you receive light pulses must decrease.

This reasoning shows that the acceleration of the elevator (a) causes a clock at (1) to tick more slowly when observed at (2). Similarly the gravitational field in elevator (b) causes a clock on the floor of the elevator at (1) to tick more slowly when observed from point (2) at the ceiling. Also a clock at the ceiling at point (2) of elevator (a) or (b) would appear to tick faster if observed at the floor at point (1).

Effects Produced by the Sun's Gravitational Field

The gravitational field of the sun causes a redshift in the sun spectrum, when observed from earth. On earth we are figuratively located at the ceiling of the elevator of Fig. 7-1 (further from the gravitational mass of the sun) and the light source is located at the floor of the elevator (on the surface of the sun). Consequently the wavelength of light emitted from the sun is greater when it is received on earth. Also a clock on the sun surface would appear to tick more slowly when observed from earth.

Assume that we could be stationed on the surface of the sun and observe a light emitted from earth. The light spectrum from the earth would be blue-shifted, and a clock on earth would appear to tick faster when observed from the sun.

Change of Spatial Dimension and Speed of Light Due to Gravity

The preceding discussion has shown how Einstein used approximate analysis to demonstrate that a gravitational field reduces a clock rate. He made other approximate analyses to demonstrate that gravity causes a spatial dimension to contract and the speed of light to decrease. These analyses of spatial contraction and speed of light are more complicated that the analysis of clock rate, and so are not discussed here.

A key conclusion from these approximate calculations is that the speed of light is affected by acceleration and gravity. This proved to Einstein that his Special Relativity theory does not apply exactly under acceleration or gravity, because Special Relativity is based on the principle that the speed of light is constant.

Einstein's Basis for Generalizing Relativity Theory

Special Relativity shows that velocity distorts measurements of distance and time. The preceding discussion shows that acceleration and gravity also distort distance and time measurements and they change the speed of light. Einstein needed to tie all of these relativistic effects together into a general theory. He found this in the complex mathematics of *tensor analysis* developed by Gregorio Ricci, which was based on a geometric principle derived by Bernhard Riemann.

In 1852, the German mathematician, Bernhard Riemann (1826-1866), presented a basic mathematical principle for characterizing curved space. Riemann derived the formulas for the *geodesic* and the *metric equation*, which describe the shortest distance between two points in curved space. Riemann was unable to develop these geometric principles in detail, because he contracted tuberculosis in 1862, and died four years later at age 39.

The Italian mathematician, Gregorio Ricci (1853-1925), used the curved-space principle of Riemann as the foundation for a comprehensive mathematical theory, which Ricci called the *absolute differential calculus*, and is now called *tensor analysis*. Tensor analysis applies calculus to curved space.

Ricci [10] published his mathematical theory in 1901 with the help of his student, Tullio Levi-Civita (1873-1941). In 1923, Levi-Civita published an updated version of this theory, an English translation of which is available as a Dover reprint. [6] *Unfortunately the monumental contributions of Ricci and Levi-Civita to General Relativity theory have generally been ignored. The mathematical foundation for General Relativity is usually referred to as Riemannian geometry. However this foundation was really the calculus of curved space developed by Ricci, which was a major extension of the Riemannian geometric principle.*

Tensor analysis has a unique formula for translating a tensor from one coordinate system to another. Therefore, by expressing Relativity principles in terms of tensors, Einstein had the basis for achieving the property called *Covariance*, in which the same laws of physics hold in all coordinate systems, regardless of velocity and acceleration, and regardless of gravitational fields. Unfortunately, tensor analysis is very complicated. It took 11 years of intensive research before Einstein was able to publish his *General theory of Relativity* in 1916 [9]. His basic Relativity theory presented in 1905 was then called the *Special theory of Relativity*.

Einstein concluded that the concept of gravitational force presented by Newton is incompatible with Relativity, because it represents a force operating instantaneously at a distance; whereas relativistic effects propagate at the speed of light. Instead, Einstein described gravity as a curvature of space. Matter causes space to be curved, and this space curvature produces the effect that we interpret as gravitational force. The Riemannian principle for specifying curved space, which is incorporated in the tensor analysis mathematics of Ricci, provided the basis for characterizing gravity in General Relativity.

Einstein formulated gravity to be consistent with relativity in the following manner. The mass of a body generates a gravitational field, which alters the motion of a second body to produce the effect of a gravitational force. When the first body moves, the change of its gravitational field propagates to the second body at the speed of light.

Einstein applied the curved-space mathematics of Riemann and Ricci to Relativity by characterizing a gravitational field as a curvature of space. Newton's theory states that a body moves in a straight line unless a force is applied to it. A straight line is the shortest distance between two points in normal (Euclidian) space. In Einstein's theory, a body moves along a geodesic, which is the shortest distance between two points in curved space, and so is equivalent to a straight line.

This curved space is four-dimensional. The orbits that the earth and a comet follow around the sun are both geodesics, but they are radically different, even though both bodies are following the shortest distance between two points. The orbits are different because the geodesics are specified in four-dimensional space-time coordinates. The earth and the comet are orbiting along paths that represent the shortest distance between two points in four-dimensional space.

Verification of General Relativity

After Einstein presented his basic (Special) theory of Relativity in 1905, he struggled intensely for 11 years to generalize his Relativity theory so that it could account for the effects of acceleration and gravity. Finally in 1916 he achieved this goal, with his General theory of Relativity, which was specified by his *gravitational field equation*.

Because of the great complexity of the Einstein equations, Einstein was only able to derive approximate solutions from them. Karl Schwartzschild (1873-1916), who was cooperating with Einstein, applied the Einstein formula to a simple physical model of a star, and thereby achieved an exact solution. Einstein also published the Schwartzschild solution in 1916. Unfortunately Karl Schwartzschild died suddenly from disease even before his famous analysis was printed. He was a German army officer at the Russian front during World War I.

The Schwartzschild solution was used as the basis for formulating tests to verify Einstein's general theory. General Relativity incorporates the Special Relativity effects caused by velocity and the General Relativity effects caused by gravity and acceleration. Within the weak gravitational fields of our solar system, the relativistic effects due to gravity are very small, but are measurable. From the Schwartzschild solution, Einstein devised the following three tests to verify the gravitational predictions of his General theory of Relativity:

(1) When a light ray passes close to the sun, it should be deflected by 1.8 arc seconds.

(2) A gravitational field causes a clock to run slower and a wavelength to increase. The gravitational field of our sun should cause the spectra of light from the sun surface to shift toward the red by 2.1 parts per million of wavelength (i.e., by 2.1×10^{-6}).

(3) The planet Mercury has a highly elliptical orbit. The axis of the Mercury orbit advances (or rotates) by 1.39 arc seconds per orbit. Of

this advance of the orbit axis, 1.29 arc seconds can be explained with Newton's laws by considering the gravitational attraction of other planets. A residual error of 0.10 arc second per orbit remained, which was explained by the Einstein General theory of Relativity.

These three tests were implemented, and the results established the validity of the Einstein General theory of Relativity.

These measurable gravitational effects of General Relativity are very small: an advance of only 0.10 arc second per orbit of Mercury; a 1.8 arc second deflection of a light beam passing close to the sun, and a gravitational redshift of only 2.1 parts per million in light emitted from the sun. Hence one might wonder why Einstein worked so hard to achieve his theory, and why General Relativity is so highly regarded. *The answer is that this generalization was essential to provide a solid theoretical foundation for the Relativity principle that is embodied in Special Relativity.*

When the predictions of General Relativity were verified, Einstein achieved great fame. After that time, Einstein did little with his General theory. Special Relativity is very much easier to apply, and has wide applicability. During Einstein's lifetime, General Relativity served primarily as a theoretical foundation for justifying Special Relativity.

Einstein could apply General Relativity only to very simple cases. With more complicated applications, the equations of General Relativity yield millions of terms, and so cannot be solved analytically. In the 1960's, a decade after Einstein's death, powerful computers became available that could apply General Relativity to complicated physical models. Since then, hundreds of scientists throughout the world have devoted their careers to computer solutions of the Einstein equations. This effort has been devoted almost exclusively to cosmology because that is about the only area that can use this expertise.

The Meaning of a Tensor

To generalize his Relativity theory, Einstein used the absolute differential calculus of Riemann and Ricci. This mathematical theory generalizes calculus so that it applies to curved space. The Riemann-Ricci mathematical theory is expressed in terms of the tensor. To see how this mathematical theory was applied to Relativity, we need to understand the meaning of a *tensor*. The *tensor* is an extension of a simpler concept called the *vector*.

The Vector

Many physical quantities have direction as well as amplitude, and so are represented by arrows, which are called "vectors". In chapter 4 we discussed the velocity of a body moving in a gravitational field. The velocity is represented by a vector arrow, where the direction of the arrow is the direction of the velocity, and the length of the arrow is proportional to the absolute value of the velocity, which is called *speed*.

Let us consider a vector that represents the velocity of an aircraft. As shown in Fig. 7-2, we can specify this velocity vector by giving the three components of airplane velocity in the easterly, northerly, and vertical directions.

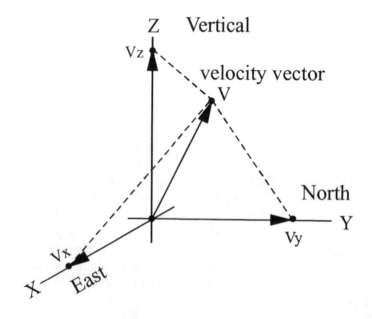

Figure 7-2: Rectangular coordinates of a velocity vector

Figure 7-2 shows three perpendicular axes, labeled x, y, z, which are called *"rectangular coordinates"*. These are also called *"Cartesian coordinates"* because they were first used by the French mathematician Rene Descartes (1596-1650). The x-axis points east, the y-axis points north, and the z-axis points in the vertical direction. Cartesian coordinates can point in other directions provided that the relative orientation of the x, y, z axes remains the same.

The vector arrow V represents the velocity of the aircraft. The three dashed lines are drawn from the tip of the velocity vector V perpendicularly to the three axes. The points where these dashed lines intersect the three axes give the x, y, z components of the vector V, which are the components of velocity in the east, north, and vertical directions. These components are measured from the intersection of the three axes (called the *origin*) along the axes to the intersection points of the three dashed lines. The components of the vector V along the x, y, z axes are denoted V_x, V_y, V_z, where V_x is the aircraft velocity in the easterly (x) direction, V_y is the aircraft velocity in the northerly (y) direction, and V_z is the aircraft velocity in the vertical (z) direction,

The Tensor

The *tensor* can be explained by considering the forces inside a mechanical body. A force exerted within a body is specified in terms of *stress*, which is the force applied per unit area. Two directions are required to describe a stress. One direction gives the direction of the force, and the second direction gives the orientation of the surface to which the force is applied. The stresses exerted within a body are specified in terms of a *tensor*.

This is illustrated in Fig. 7-3, which shows the forces applied to a cube of material within a body. Each face of the cube is assumed to have unit area. *Force-per-unit-area* is called *stress*, and so the forces applied to the unit-area faces of the cube are stresses. The symbol "p" is used to denote a stress, because pressure is a typical example of a stress.

The stresses in Fig. 7-3 are expressed in term of the x, y, z axes. These stresses are denoted in the form p_{ab}, where the subscript indices a, b can each represent x, y, or z. The first subscript index (a) describes the direction of the force. The second index (b) describes the orientation of the face to which the force is applied. The orientation of a face is defined by a vector that is perpendicular to the face.

Diagram (a) shows the three *compression stresses* that are applied at three faces of the cube, which are denoted p_{xx}, p_{yy}, and p_{zz}. Stress p_{xx} is exerted on the cube in the x-direction at a face that is perpendicular to the x-direction. Stresses p_{yy} and p_{zz} are defined in a similar manner.

The three faces of the cube, which are specified by the second subscript index, are labeled the *x-face*, the *y-face*, and the *z-face*. The x-face is perpendicular to the x-direction, the y-face is perpendicular to the y-direction, and the z-face is perpendicular to the z-direction.

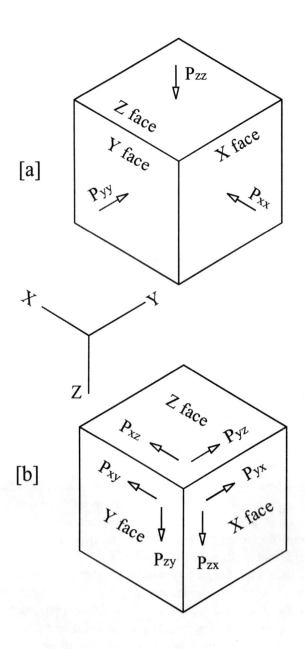

Figure 7-3: Internal stresses within a mechanical body: [a] compression stresses; [b] shear stresses.

Diagram (b) of Fig 7-3 shows the six *shear stresses*, which are applied parallel to the cube faces. The two shear stresses applied to the x-face are denoted p_{yx} and p_{zx}. Stress p_{yx} is applied to the x-face in the y-direction, and stress p_{zx} is applied to the x-face in the z-direction. Similarly, the two shear stresses applied to the y-face are denoted p_{xy} and p_{zy}, and the two shear stresses applied to the z-face are denoted p_{xz} and p_{yz}.

Therefore nine separate stresses are required to describe the internal forces within a mechanical body. These stresses are arranged as follows in a 3-by-3 array, which is called a *matrix*

$$\begin{vmatrix} p_{xx} & p_{xy} & p_{xz} \\ p_{yx} & p_{yy} & p_{yz} \\ p_{zx} & p_{zy} & p_{zz} \end{vmatrix}$$

Elements are arranged in this matrix such that the first index indicates the row, and the second index indicates the column. For example, in the first row the first index is x; in the second row the first index is y; and in the third row the first index is z.

This matrix is defined by the variable p_{ab}, which is called a *tensor*. This variable p_{ab} represents the nine stresses in a general form. The subscripts a and b are called indices, and each index can represent x, y, or z.

In summary, a tensor with two indices is required to define a stress, because stress has two independent directions. The first index specifies the direction of the force, and the second index specifies the orientation of the surface to which the force is applied. A surface is perpendicular to the direction specified by the second index.

It is often convenient to describe the x, y, z coordinates by numbers, where x is called x_1, y is called x_2, and z is called x_3. Hence the x index of a stress component is replaced by 1, the y index is replaced by 2, and the z index is replaced by 3. In this form the stress tensor matrix is expressed as

$$\begin{vmatrix} p_{11} & p_{12} & p_{13} \\ p_{21} & p_{22} & p_{23} \\ p_{31} & p_{32} & p_{33} \end{vmatrix}$$

Tensors in Relativity Theory. In Relativity theory, the space and time variables cannot be considered separately. Reality is described by *events*, where an event is specified by three spatial coordinates plus a time coordinate, and so has four dimensions. In Relativity theory a vector has 4 components, and a tensor has 4x4 or 16 components.

In a Relativity tensor, the x, y, z spatial coordinates are specified as x_1, x_2, x_3. Einstein denoted the time coordinate as x_4, but this book uses the alternative convention where the time coordinate is denoted x_0. Hence a Relativity tensor is represented by a 4-by-4 array of the form.

$$\begin{vmatrix} p_{00} & p_{01} & p_{02} & p_{03} \\ p_{10} & p_{11} & p_{12} & p_{13} \\ p_{20} & p_{21} & p_{22} & p_{23} \\ p_{30} & p_{31} & p_{32} & p_{33} \end{vmatrix}$$

The first row and column give the tensor elements associated with time.

Application of the Tensor to General Relativity

Let us see how tensors are applied in the Einstein General theory of Relativity. The major tensors used in the Einstein theory are:

metric tensor (g_{ab})
Ricci tensor (R_{ab})
energy-momentum tensor (T_{ab})

The *metric tensor* specifies the shortest distance between two points in curved space. The *Ricci tensor* specifies the curvature of space. If a region of space has no gravitational field, and hence no curvature, all elements of the Ricci tensor are zero. The *energy-momentum tensor* specifies the characteristics of matter and electromagnetic fields.

The basic steps for solving the Einstein theory are outlined in Table 7-2. In step (1) the extremely complicated equations of the general Riemann-Ricci mathematical theory are applied to calculate the Ricci tensor from the metric tensor. Step (2) is the Einstein gravitational field equation, which relates the Ricci tensor to the energy-momentum tensor. Step (3) calculates the energy-momentum tensor from the characteristics of the physical model. Let us examine the tensors of this calculation

7. Einstein General Theory of Relativity

Table 7-2: Outline of steps in Einstein theory calculation

(1) metric tensor ⇒ Ricci tensor
(2) Ricci tensor ⇔ energy-momentum tensor
(3) energy-momentum tensor ⇐ physical model

(1) Riemann-Ricci mathematical theory
(2) Einstein gravitational field equation
(3) General formulas for calculating energy-momentum tensor

The Metric Tensor

The metric tensor describes the shortest distance between two points in curved space. To illustrate this concept, let us apply it in two dimensions by considering points on the curved surface of the earth. Boston, Massachusetts and Rome, Italy are at approximately the same latitude. If an airplane is flying from Boston to Rome, one might think that the airplane would fly directly east, but this is not so. That is the long way. The airplane follows a great-circle route, which is the shortest distance between these two cities. The airplane initially flies nearly north-east and ends by flying nearly south-east. The great-circle route is obtained by passing a plane through the center of the earth, Boston, and Rome, and finding the intersection of the plane with the earth's surface.

A metric tensor could be specified in two dimensions to describe the curved surface of the earth. This metric tensor would give the great-circle route, which is the shortest distance between two points. In Relativity theory, the metric tensor is specified in four dimensions, and gives the shortest distance between two points in four-dimensional space-time coordinates.

In its general form, the metric tensor has 16 elements as follows:

$$\begin{vmatrix} g_{00} & g_{01} & g_{02} & g_{03} \\ g_{10} & g_{11} & g_{12} & g_{13} \\ g_{20} & g_{21} & g_{22} & g_{23} \\ g_{30} & g_{31} & g_{32} & g_{33} \end{vmatrix}$$

The elements along the diagonal of the tensor (g_{00}, g_{11}, g_{22}, g_{33}) are called the diagonal elements of the tensor, and the other elements are the nondiagonal elements.

When all of the nondiagonal elements are zero, the tensor is called a "diagonal tensor". For the cases that Einstein could analyze, the metric tensor was always diagonal, and had the following form:

$$\begin{vmatrix} g_{00} & 0 & 0 & 0 \\ 0 & g_{11} & 0 & 0 \\ 0 & 0 & g_{22} & 0 \\ 0 & 0 & 0 & g_{33} \end{vmatrix}$$

A diagonal metric tensor has only four elements, (g_{00}, g_{11}, g_{22}, g_{33}.). The element g_{00} represents the time-component of the metric tensor and g_{11}, g_{22}, g_{33} represent the three-dimensional spatial components.

The Einstein theory can be solved analytically only when the metric tensor is diagonal. If the metric tensor is not diagonal, the Riemann-Ricci equations relating the metric tensor to the Ricci tensor yield millions of terms, and so can only be solved on a computer. They cannot be solved analytically. *Since powerful computers were not available during Einstein's lifetime, Einstein had to limit the application of his General Relativity theory to very simple physical models that could yield diagonal metric tensors.*

In the stress tensor illustrated in Fig. 7-3, the diagonal stress elements are the compression stresses (diagram a), and the nondiagonal elements are the shear stresses (diagram b).

The Ricci Curvature Tensor

The Ricci tensor describes the curvature of space. If a region of space has no gravitational field, and hence no curvature, all elements of the Ricci tensor R_{ab} are zero.

The Riemann-Ricci mathematical theory has a precise but very complicated formula that allows one to calculate the Ricci tensor R_{ab} from the metric tensor g_{ab}. If the metric tensor is diagonal, it is possible to calculate analytically the elements of the Ricci tensor from those of the metric tensor. However, it may take a skilled mathematician several weeks to perform this calculation analytically without error. If the metric tensor is not diagonal, this formula yields million of terms, and so can only be implemented on a computer.

The Energy-Momentum Tensor

The energy-momentum tensor describes the stress and energy of matter and electromagnetic fields. There are precise formulas for calculating the energy-momentum tensor T_{ab} from physical models of matter and electromagnetic radiation.

The Einstein Gravitational Field Equation

The Einstein gravitational field equation is a tensor formula that relates the Ricci tensor to the energy-momentum tensor. This formula represents 16 separate equations. However, the tensors of the Einstein theory are symmetric. For example g_{21} is equal to g_{12}. Because of this symmetry, 6 of the 16 equations of the Einstein gravitational field equation are redundant. Hence this formula represents 10 independent equations. ***Thus, in its general form, the Einstein gravitational field equation represents 10 independent equations that relate the Ricci tensor to the energy momentum tensor.***

Outline of the Einstein Theory Calculations

Table 7-2 showed the major steps involved in applying the Einstein theory. In step (3) the energy-momentum tensor is calculated from the characteristics of the physical model using standard formulas. In step (2) the Einstein gravitational field equation is used to calculate the Ricci tensor from the energy-momentum tensor.

The goal of the analysis is to derive the metric tensor that will yield the required Ricci tensor. However, the complicated calculations of step (1) can be solved only in a single direction. They allow one to compute the Ricci tensor from the metric tensor. To apply the Einstein theory, these extremely complicated equations must be solved backward. Achieving this backward calculation is a formidable mathematical task.

The above discussion has outlined the tensors used in the Einstein theory. To see how these tensors are used in solving the Einstein theory. let us examine the famous Schwartzschild solution. At the time that Einstein presented his gravitational field equation in 1916, the only exact solution of his theory was the one obtained by Schwartzschild.

The Schwartzschild Solution of the Einstein Theory

To achieve a solution to the complicated Einstein equations, Schwartzschild assumed a very simple model for a star, which could be our sun. He assumed that the star had no viscosity and had a constant density of matter. Assuming no viscosity eliminated the non-diagonal shear stresses (illustrated in Fig 7-3), and this yielded an energy-momentum tensor that was diagonal. Since a star is gaseous, the assumption of constant mass density is highly inaccurate, but was needed to achieve a solution.

The surface of the sun has about the same density as the earth's atmosphere, which is 1/1000 of the density of water. The center of the sun has 100 times the density of water. Hence the sun's density varies with radius by a factor of 100,000. Nevertheless, the Schwartzschild analysis approximated the sun as a fluid having a constant density of matter. Although this approximation was very unrealistic, it was essential. Without it, the Einstein equations could not be solved analytically.

With standard formulas, Schwartzschild calculated the energy-momentum tensor for his simple physical model of a star. He then used the Einstein gravitational field equation to calculate the Ricci tensor from the energy-momentum tensor.

There are very complicated equations expressing the Ricci tensor R_{ab} in terms of the metric tensor g_{ab}. Schwartzschild had to solve these equations *backward* to find the metric tensor g_{ab} that would yield his required Ricci tensor. He achieved his backward calculation by assuming general formulas for the metric tensor elements having unknown parameters. From these general formulas he calculated the corresponding formulas for the elements of the Ricci tensor. He set these general formulas equal to the values for the Ricci tensor that he had calculated, and solved the resultant equations. This gave him the unknown parameters for the elements of the metric tensor, and thereby gave him complete equations for the metric tensor elements.

There are two separate solutions in the Schwartzschild analysis. The above discussion pertains to the *interior solution*, which applies inside the star. The *exterior solution* applies in the vacuum of space outside the star. In a vacuum, all elements of the energy-momentum tensor are zero, and so the Einstein gravitational field equation shows that all elements of the Ricci tensor must be zero in a vacuum.

Thus Schwartzschild had two separate solutions, his *interior solution* applied inside the star, and his *exterior solution* applied outside the star. He matched his two solutions at the surface of the star, and thereby achieved his final result.

The Schwartzschild analysis started with physical models for a star and the vacuum of space outside the star. The energy momentum tensors were obtained from these models. From the energy-momentum tensors, Schwartzschild calculated the corresponding metric tensors. When the metric tensor is known, one can calculate the relativistic effects produced by gravity. For example, the change in clock rate produced by gravity is proportional to the square root of g_{00}. The calculation of relativistic effects from the metric tensor is discussed in Chapter 9.

Computer Solutions of the Einstein Theory

The Einstein General theory of Relativity can be solved analytically only when the metric tensor is diagonal. If the metric tensor is not diagonal, the calculations yield millions of terms. During Einstein's lifetime, the theory could be applied only to very simple physical models that resulted in diagonal metric tensors.

In the 1960's, computers became available that could calculate these millions of terms and so could solve the Einstein gravitational field equation for a complex physical model. However, even with a powerful computer the solution of the Einstein gravitational field equation for a general model is very difficult, because the equations yield millions of terms and must be solved backward. A sophisticated iterative computer program is needed to achieve this backward calculation.

This iterative program starts with approximate values for the metric tensor elements, and from these it computes the corresponding elements of the energy-momentum tensor. The computed elements of the energy-momentum tensor are compared with the desired elements, and the differences are used to change the metric tensor elements in such a way as to minimize the differences. The program cycles through this process until the computed elements of the energy-momentum tensor match the desired elements.

The iterative program may implement billions of cycles before a solution is obtained. Sophisticated mathematical algorithms have been developed to achieve iterative computer programs that converge to solutions. Since the 1970's, many hundreds of scientists have devoted their careers to the task of solving the formidable Einstein equations on the computer. Highly advanced analytical procedures have been devised to achieve these computer solutions.

Scientists applying the Einstein theory have often been called "geniuses like Einstein". Einstein displayed a profound genius in developing General Relativity, but a similar intellect is not needed to apply this theory. The mathematical formulas associated with the Einstein theory are precisely specified. Although it takes great mathematical skill to solve these equations on a computer, these computational skills do not place a theoretician solving the Einstein equations in the same intellectual category as Albert Einstein.

Chapter 8

The Yilmaz Theory of Gravity

Derivation of the Yilmaz Solution

As was shown in Table 7-1 of Chapter 7, Einstein proved that when light rises against a gravitational field it experiences a $\Delta\lambda/\lambda$ redshift approximately equal to (gh/c^2), where g is the acceleration of gravity and h is the height through which the light rises. While performing his PhD research at the Massachusetts Institute of Technology in the early 1950's, Huseyin Yilmaz examined this approximate calculation by Einstein, and discovered that he could solve the problem exactly. This exact calculation of redshift yielded an exact value for the metric tensor element g_{00}, which characterizes time. In performing this analysis, Yilmaz applied principles of Special Relativity that had been developed by Einstein.

Yilmaz then postulated that the speed of light measured locally in a gravitational field should be independent of direction. He proved that for this condition to hold the following must apply:

$$g_{11} = -1/g_{00} \; ; \; g_{11} = g_{22} = g_{33}$$

With these relations, Yilmaz calculated exact values for the metric tensor elements g_{11}, g_{22}, g_{33}, and thereby achieved an exact solution to the principles of General Relativity, which resulted in the Yilmaz theory of gravity. The derivation of the metric tensor elements for the Yilmaz theory is given in *Story* [4] (Appendix E).

Yilmaz mailed his results to Einstein. However, Albert Einstein was too sick to read them, and died soon thereafter.

8. The Yilmaz Theory of Gravity

The Yilmaz Gravitational Field Equation

With his exact solution, Yilmaz derived in a rigorous manner the corresponding gravitational field equation. The gravitational field equation for the Einstein theory relates the Ricci tensor to the energy-momentum tensor. In addition to the Ricci tensor and the energy-momentum tensor, the gravitational field equation for the Yilmaz theory has a third tensor that represents the energy of the gravitational field. The fact that the Einstein theory lacks a tensor to specify the energy of the gravitational field is a serious limitation.

As shown in *Story* [4] (Appendix E), Einstein searched for a tensor to specify the energy of the gravitational field, but was unable to derive a true tensor for this purpose. He was only able to derive a *"pseudo-tensor"*, which he believed represented the *"energy components of the gravitational field"*. Since this was not a true tensor, Einstein could not use it in his gravitational field equation.

Einstein obtained his gravitational field equation in an intuitive manner, after many years of searching for a solution. His equation appeared to work, and so he and other scientists believed that he had found a rigorous specification of Relativity principles. However, this equation has only been verified by experiments performed within our solar system, where the relativistic effects produced by gravity are extremely small.

In contrast, Yilmaz derived his gravitational field equation from rigorous analysis. Within the weak gravitational fields of our solar system, the Einstein and Yilmaz theories yield essentially the same predictions. However, in an intense gravitational field the theories predict radically different results. The Yilmaz theory does not allow the physically impossible singularity solutions that have been obtained from the Einstein theory.

With its third tensor representing the energy in the gravitational field, the Yilmaz gravitational field equation is more complicated than that of the Einstein theory. However, Yilmaz has a general solution to his gravitational field equation, and so the Yilmaz gravitational field equation is not used in practical calculations. With its general solution, the Yilmaz theory is very much easier to apply than the Einstein theory.

The General Time-Varying Yilmaz Theory

The initial theory derived by Yilmaz applies exactly only when the gravitational field does not vary with time, and so is called a *static solution*. (The Schwartzschild solution of the Einstein theory is also a *static solution*.) The first paper on the Yilmaz theory, which presented this static solution, was published in 1958. After 15 years of intense research, Yilmaz was able to generalize his theory to obtain his *time-varying solution*, which was published in 1973. [Y3]

A derivation of the time-varying Yilmaz theory is described in the *Website* [3], *Addendum* page 5, Chapter 5 and Appendix F. The time-varying Yilmaz theory is much more complicated than the static solution, yet is still very much easier to apply than the Einstein theory.

One can prove from the general time-varying Yilmaz theory that the simple static solution of the Yilmaz theory gives a very accurate approximation if the gravitational field varies slowly relative to the speed of light. This condition is satisfied in nearly all practical applications, and so the time-varying solution is rarely needed

The time varying solution of the Yilmaz theory is needed to describe gravitational waves. (Gravitational waves have also been predicted by the Einstein theory.) Except for gravitational waves, the simple static solution of the Yilmaz theory is nearly always more than adequate.

Discussion of the Yilmaz Theory

The Yilmaz gravitational theory is a direct extension of the Einstein General theory of Relativity, which applies the principles that Einstein specified in developing his theory. Hence the Yilmaz theory is a refinement of the Einstein theory.

Yilmaz discovered an exact solution to the relativistic principles that Einstein established. If Einstein had retraced his steps, he might have discovered the exact solution that Yilmaz found. This exact solution would have been far more desirable than the gravitational field equation that Einstein chose to specify his theory. Einstein developed his gravitational field equation intuitively, after many years of searching for an answer.

The derivation of the Einstein gravitational field equation is explained by John A. Peacock in his lengthy scientific book, titled *Cosmological Physics* [22]. This book describes theoretical Big Bang research and includes a detailed discussion of Relativity theory.

Peacock states [22] (p. 19) that the Einstein gravitational field equation *"cannot be derived in any rigorous sense; all that can be done is to follow Einstein and start by thinking about the simplest form such an equation might take."*

On pp. 26-27, Peacock [22] discusses possible *"alternative theories* of gravity". He asks, "How certain can we be that Einstein's theory of gravitation is correct?" He notes that the Einstein theory does not apply in sub-atomic scales, and goes on to say, "Apart from this restriction, there are no obvious areas of incompleteness. . . Nevertheless, . . it is possible that more accurate experiments will yield discrepancies. Over the years this possibility has motivated many suggestions of alternatives to general relativity."

Peacock is a strong supporter of the Big Bang theory, which is solidly tied to the Einstein gravitational field equation. Nevertheless, even Peacock admits that the Einstein gravitational field equation was obtained in an intuitive manner, and that a number of alternatives to that equation have been proposed by responsible scientists.

In contrast, Yilmaz derived the gravitational field equation of his theory by means of rigorous analyses that applied the principles of the Einstein theory. The static solution of the Yilmaz theory is remarkably simple even though the theory has a very solid mathematical foundation. The simple static solution is more than adequate for nearly all applications.

The Yilmaz theory does not allow the physically impossible black hole and Big Bang singularities that have been derived from the Einstein gravitational field equation. Einstein strongly opposed the singularity solutions obtained from his theory.

Reason for Opposition to the Yilmaz Theory

With these great advantages of the Yilmaz theory, one might think that the scientific community would clamor for it. This might be true if the Einstein General theory of Relativity were being used in practical applications, but that is very rarely the case.

The Einstein theory is the basis for an enormous amount of cosmological research performed over the past 30 years by hundreds of scientists throughout the world. Computer studies of the Einstein gravitational field equation are the foundation for this research. Because of the extreme mathematical complexity of the Einstein gravitational field equation, sophisticated mathematical techniques have been required to analyze it on the computer.

If the validity of the Yilmaz theory were accepted, it would be recognized that the Einstein gravitational field equation is flawed. This finding would make obsolete and irrelevant the enormous computer research effort that has been based on the Einstein equation. The sophisticated computer techniques are not needed to apply the simple

Yilmaz theory. Besides, the Yilmaz theory refutes the singularity predictions that have been derived from this computer research.

Therefore, it is not surprising that the army of General Relativity experts that are involved in Big Bang cosmology studies are strongly opposed to the Yilmaz theory.

Consistency with Quantum Mechanics

After presenting his General theory of Relativity, Einstein did little with this theory. He devoted the rest of his life primarily to the task of developing a *unified field theory*. This would have combined into a single theory the concepts of gravitational fields, electromagnetic fields, and atomic nuclear fields. He never succeeded, although he struggled with this task until his last days. An important reason for his failure is that the Einstein gravitational field equation conflicts with quantum mechanics.

Yilmaz has proven that his gravitational field equation is consistent with quantum mechanics. This finding opens great possibilities for relating Relativity theory to quantum field theory and may eventually lead to Einstein's elusive goal of a *unified field theory*.

Diagonal Character of Yilmaz Metric Tensor

The metric tensor for the Einstein theory is diagonal only for very simple physical models. In contrast, the metric tensor for the static solution of the Yilmaz theory is always diagonal, and so consists only of the four elements: g_{00}, g_{11}, g_{22}, and g_{33}. This condition holds even for very complicated physical models, provided that the gravitational field changes at velocities much less than the speed of light, a condition that is almost always satisfied.

The fact that the metric tensor for the Yilmaz theory is almost always diagonal for practical applications indicates that the Yilmaz theory is remarkably easy to apply.

Chapter 9

Applying the Einstein and Yilmaz Theories

Relativistic Effects Produced by Gravity

This chapter applies the Einstein and Yilmaz theories to obtain plots of the relativistic effects produced by the gravitational field of a star.

Special Relativity shows that velocity causes a clock to run slower and a measuring rod to contract. In General Relativity, a gravitational field also causes a clock to run slower and a measuring rod to contract. In Special Relativity the speed of light is always constant, but in General Relativity a gravitational field causes the speed of light to decrease. Because of the reduction of clock rate, excited atoms oscillate at lower frequency, and this produces a gravitational redshift.

As shown in Table 9-1, these gravitational effects can be calculated from the metric tensor elements. These formulas are obtained from *Story* [4], Chapter 10.

Table 9-1: *General formulas for relativistic effects due to gravity*

clock rate:	$\sqrt{g_{00}}$	speed of light:	$\sqrt{g_{00}/(-g_{11})}$
wavelength:	$1/\sqrt{g_{00}}$	spatial contraction:	$1/\sqrt{-g_{11}}$

The following discussion applies the Einstein and Yilmaz theories to specify the gravitational effects of a star. The Einstein theory uses the Schwartzschild solution, which requires that the star have a constant density of matter. With the Yilmaz solution, the density of the star can vary with radius, and so the Yilmaz analysis applies to a very much more realistic physical model of a star than does the Schwartzschild solution.

The only restriction to this Yilmaz analysis is that the density of the star must be spherically symmetric.

The Normalized Relativistic Mass m

The gravitational field of a star is proportional to the ratio M/r, where M is the mass of the star and r is the radial distance from the center of the star. To handle the M/r ratio conveniently, a normalized mass (m) is defined as (MG/c^2), as shown in Table 9-2.

Table 9-2: *Calculation of normalized mass m of sun, and the m/r ratio at the surface of the sun*

(1)	Normalized mass m	MG/c^2
(2)	Mass of sun M	2×10^{30} kilogram
(3)	Normalized mass m of sun	1.5 kilometer (km)
(4)	Radius of sun	700,000 km
(5)	m/r ratio at sun's surface	(1.5 km)/(700,000 km) = 2.1×10^{-6}

In Table 9-2, item (1) gives the general formula for normalized mass. Item (2) gives the mass M of the sun, which is 2×10^{20} kilogram. The number 2×10^{20} means 2 followed by 20 zeros. The value of the gravitational constant G in item (1) is given in the Glossary, and the speed of light c is 300,000 km/sec. Applying these numbers to the formula of item (1) shows that the normalized mass m of the sun is 1.5 kilometers (km), as given in item (3).

Normalized mass has units of distance (kilometers). Normalized mass is a convenient artifice that simplifies our discussion; it does not imply that true mass can be measured in kilometers.

In Table 9-2, item (4) gives the radius of our sun, which is 700,000 kilometers. Item (5) gives the m/r ratio at the surface of the sun, which is obtained by dividing the normalized sun mass m (1.5 km) of item (3) by the sun's radius (700,000 km) of item (4). The m/r ratio at the sun's surface is 2.1×10^{-6}. As shown in the glossary, 10^{-6} means $1/10^6$, which is 1/1,000,000. Thus, 2.1×10^{-6} means 2.1 parts per million, or 0.000,0021. Since m and r have the same distance units (kilometers), the m/r ratio is a simple number without units.

The m/r ratio describes the gravitational field of the sun or of any other body. In our solar system, the maximum value of the m/r ratio occurs at the surface of the sun, where it is 2.1 parts per million. The radius of the earth's orbit is 150 million km. Therefore the m/r ratio at the orbit of the planet earth is (1.5 km)/(150 million km), which is one part in 100 million (10^{-8}).

This shows that the m/r ratio describing the gravitational field of our sun is a very small number. *The maximum (m/r) ratio in our solar system occurs at the surface of our sun, where it is 2.1 parts per million. At the distance of the earth orbit, the m/r ratio is only one part in 100 million.*

Effect of Gravity on the Speed of Light

Figure 9-1 shows how the speed of light varies with the gravitational field of a star is accordance with the Einstein and Yilmaz theories. The formulas for these plots and the plots in Figs. 9-2 and 9-3 are obtained from *Story* [4], Chapter 10. They apply the general formulas given in Table 9-1. The plots are shown as solid curves for the Schwartzschild solution of the Einstein theory, and as dashed curves for the Yilmaz solution.

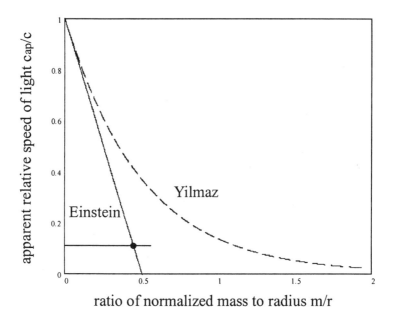

Figure 9-1: Apparent relative speed of light for Schwartzschild Einstein solution and for Yilmaz solution.

In Fig. 9-1, the horizontal axis gives the gravitational field of the star expressed in terms of the m/r ratio at the surface of the star. Since the m/r ratio at the surface of our sun is 2.1 parts per million, only the first tiny bit of this plot applies to our sun.

The vertical axis shows the apparent speed of light c_{ap} divided by the normal speed of light c experienced on earth (300,000 km/sec). The plot starts with a ratio of unity, which means that at zero gravitational field the apparent speed of light is equal to the normal speed of light. Within our solar system, the m/r ratio on the horizontal axis is extremely small, and so the apparent speed of light c_{ap} is very close to the normal speed of light c.

The plots of Fig. 9-1 are significant when we consider an extremely dense star, such as a neutron star. The solid plot for the Einstein theory drops to zero when the m/r ratio reaches 0.5. This value (0.5) is 240,000 times greater than the m/r ratio at the sun's surface (2.1×10^{-6}).

If a star with 8 times the mass of our sun should collapse to the density of a neutron star, it would have an m/r ratio of 0.5 at its surface. Therefore, the Einstein theory predicts that the speed of light should be zero at the surface of this neutron star. In contrast, the speed of light for the Yilmaz theory never goes to zero at any m/r value.

The Black Hole Singularity

The plot for the Einstein theory in Fig 9-1 is derived from the Schwartzschild solution. If the m/r ratio exceeds 0.5, the Schwartzschild solution predicts that pressure inside the star should become "imaginary", which is an impossible result. This means that the Schwartzschild analysis does not yield an answer when m/r exceeds 0.5. The value 0.5 is called the *Schwartzschild limit*, which represents the upper limit for the m/r ratio in the Schwartzschild analysis.

At the point where m/r is equal to 0.5, the radius r is equal to 2m. This distance (2m) is called the *Schwartzschild radius*. For our sun the Schwartzschild radius is 3 kilometers. If the Schwartzschild radius lies inside the star (as it does for our sun), the Schwartzschild analysis applies. If a star is so compact that the Schwartzschild radius lies outside the star, the Schwartzschild analysis does not yield a solution.

As was explained earlier, Oppenheimer and Snyder in 1939 showed that they could obtain a solution from the Einstein equations when m/r exceeds the Schwartzschild limit, if they assume that the radius of the star decreases with time. Therefore Oppenheimer and Snyder concluded from their analysis that if m/r exceeds the Schwartzschild limit of 0.5, the star must contract "indefinitely".

There is no limit in this contraction process, and so they concluded that the star should shrink until it becomes a *singularity* having a diameter of zero. Since the star's mass does not change as the star shrinks, the density of matter should become infinite.

Einstein strongly opposed this *Schwartzschild singularity*. In his rebuttal he insisted that, **"Schwartzschild singularities do not exist in physical reality"**. Einstein required his theory to be consistent with experimental evidence, and he realized that a singularity would drastically violate such evidence. Einstein presented an analysis to show that the predicted *Schwartzschild singularity* would not actually occur.

While Einstein was alive, no scientist attempted to dispute Einstein's refutation of the *Schwartzschild singularity*. Nevertheless, the concept became a popular notion in science fiction. If the Schwartzschild limit can be exceeded, the star would be surrounded by a spherical surface over which the speed of light is zero. The surface was called the ***event horizon***, because time would theoretically stand still over this surface. The event horizon surface would be a sphere with the Schwartzschild radius ($r = 2m$).

Light theoretically cannot escape from inside the event horizon surface, and so the star was called a *"black hole"*. There were many science fiction accounts of the concept of *"falling into a black hole"*. The stories claimed that the gravitational force is so great that nothing can oppose the tremendous gravitational pull of a *black hole*.

Although the black hole concept is well known to the general public, it is not generally recognized that the star inside a black hole must contract "indefinitely" until it shrinks to form a singularity having zero diameter and an infinite density of matter.

The dashed plot for the Yilmaz theory in Fig. 9-1 shows that the Yilmaz theory does not allow a black hole. The speed of light never goes to zero, for any m/r ratio. The Yilmaz theory shows that the black hole concept is nothing more than a mathematical defect in the Einstein gravitational field equation. Since the Yilmaz theory is a refinement of the Einstein theory, the Yilmaz theory has proven that the basic principles of the Einstein General theory of Relativity are inconsistent with the black hole. ***In agreement with Einstein, Yilmaz has proven that the basic Einstein theory, when properly refined, does not predict a physically impossible singularity.***

When computer studies of the Einstein gravitational field equation were implemented in the 1960's, about a decade after Einstein's death, it was concluded that the Einstein theory does indeed predict a singularity.

Consequently the black hole concept was officially endorsed by the scientific community It was widely believed that a black hole must actually exist "in physical reality".

Misleading Astronomical Evidence for Black Holes

In recent years, many astronomers have made observations that they claim are proof of the existence of black holes. But how does one detect a black hole? A black hole does not emit radiation to indicate its existence. What astronomers are observing are massive, highly compact bodies. The obvious interpretation is that these bodies are massive neutron stars.

Since the Einstein theory does not allow a neutron star with a mass-to-radius ratio that exceeds the Schwartzschild limit, astronomers conclude that these massive compact bodies cannot be massive neutron stars. They must be black holes.

The answer to this enigma is that the Yilmaz theory should be applied, not the Einstein theory. The Yilmaz theory allows a massive neutron star, and can readily explain all of these astronomical observations as neutron stars.

A neutron star, with its extreme density of one billion tons per teaspoon, is a very strange star, but is physically real. In contrast, the star at the center of a black hole is a physically impossible singularity having zero size and an infinite density of matter. Einstein absolutely rejected the black hole singularity.

Second Limit to the Schwartzschild Solution

Although the Schwartzschild analysis yields a real solution up to the Schwartzschild limit where m/r is equal to ½, a physically impossible condition theoretically occurs at a lower value of m/r. If m/r is greater than 4/9 but less than ½, the Schwartzschild analysis predicts that the pressure inside the star should be infinite over a spherical surface. This surface should progress from the center of the star to the circumference as the m/r ratio is increased from 4/9 to ½.

Since infinite pressure is not physically possible, it has been concluded that a star would become unstable if m/r exceeds 4/9, and would collapse to form a black hole. The dot on the plot for the Einstein theory in Fig. 9-1 shows this upper limit to the m/r ratio. It is generally assumed that a star cannot be stable if m/r exceeds this value. If m/r is greater than 4/9 the star must collapse to become a black hole. As a star collapses, the m/r ratio at its surface increases, and so the Schwartzschild limit of 0.5 is quickly exceeded.

Gravitational Effects on Distance and Clock Rate

A gravitational field causes a distance to contract and a clock to run slower. Figure 9-2 shows the contraction of dimension and the reduction of clock rate that are predicted by the Einstein and Yilmaz theories. For these analyses, the plots of relative clock rate and spatial contraction ratio are equal, and so can be expressed as a single plot.

As in Fig. 9-1, the horizontal scale gives the m/r ratio. The vertical scale shows the contraction of dimensions and the reduction of clock rate that are produced by a gravitational field. According to the Einstein theory, a dimension or a clock rate decreases from its normal value (for m/r equal to zero) to a value of zero when the m/r ratio reaches the Schwartzschild limit of 0.5.

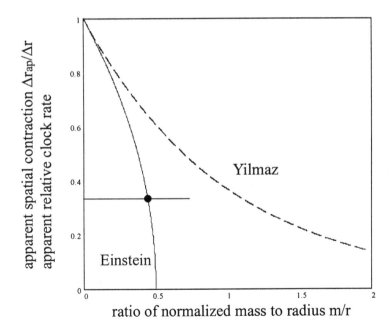

Figure 9-2: Apparent spatial contraction and relative clock rate for Einstein Schwartzschild solution and Yilmaz theory

On the other hand, we have seen that if m/r exceeds 4/9, the Einstein theory predicts that the star should become unstable, and should collapse to become a black hole. Therefore the maximum value for m/r that is allowable by the Einstein theory for a stable star is 4/9. This limit is shown by a dot on the plot for the Einstein theory in Fig. 9-2. At this point, the value of spatial contraction or relative clock rate is 1/3.

According to the Einstein theory, Fig. 9-2 shows that an intense gravitational field can reduce a spatial dimension or a clock rate to 33.3 percent of its normal value. A star that might have a sufficient m/r ratio to achieve greater reduction of dimensions or clock rate should become unstable and collapse to become a black hole. After a star becomes a black hole it does not radiate, and so its spatial dimensions and clock rate are unobservable.

The dashed curve for the Yilmaz theory in Fig. 9-2 does not have a lower limit. Therefore, the Yilmaz theory shows that there is no lower limit to the spatial contraction or relative clock rate that can be produced by a gravitational field.

Effect of Gravity on Wavelength

The decrease of clock rate shown in Fig. 9-2 indicates that gravity causes a clock period to increase. Since the wavelength of a light wave is equivalent to a clock period, a gravitational field causes wavelength to increase by an amount inversely proportional to the decrease of clock rate. Figure 9-3 shows how wavelength changes with m/r ratio.

Figure 9-3 gives the wavelength ratio λ'/λ, where λ is the normal wavelength and λ' is the observed wavelength. Since the maximum allowable value of m/r for the Einstein theory is 4/9, Fig. 9-3 shows that the maximum value of the λ'/λ wavelength ratio for the Einstein theory is 3. This maximum value is shown as a dot in the figure.

The observed wavelength λ' is equal to $(\lambda + \Delta\lambda)$, where $\Delta\lambda$ is the increase in wavelength above the normal wavelength λ. Redshift is defined as the ratio $\Delta\lambda/\lambda$ and so is equal to $(\lambda'/\lambda - 1)$. Figure 9-3 shows that the maximum λ'/λ wavelength ratio due to gravity that can be predicted by the Einstein theory is 3, and so the maximum $\Delta\lambda/\lambda$ gravitational redshift that the Einstein theory can predict is 2.

In Fig. 9-3, the dashed curve of λ'/λ for the Yilmaz theory does not have an upper limit. Therefore the Yilmaz theory shows that there is no upper limit to the value of gravitational redshift.

Chapter 3 discussed the quasar, which is a strange star with an extremely large redshift in its spectrum. It is generally assumed that this redshift is produced by velocity, which would indicate that quasars are receding at velocities close to the speed of light. However, Fig. 9-3 shows that an intense gravitational field can also produce a large redshift, and might explain the anomalous redshift of the quasar.

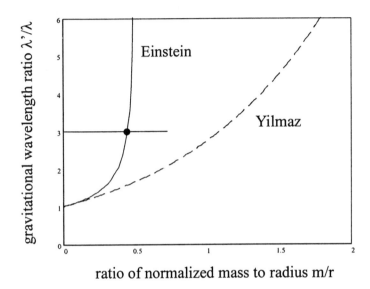

Figure 9-3: Wavelength ratio due to gravity for Einstein Schwartzschild solution and for Yilmaz theory

Chapter 10

Evidence Against the Big Bang

For many years the field of astronomy has been treating the Big Bang theory as fact, even though there are strong reasons to doubt the theory. Nevertheless, many voices have been raised against the Big Bang. Let us consider some of them.

The Editorial of Geoffrey Burbidge

The case against the Big Bang was expressed eloquently by Professor Geoffrey Burbidge in an editorial article of the February 1992 *Scientific American* [25]. Professor Burbidge is the former director of the Kitt Peak National Observatory, and is presently Professor of Astrophysics at the University of California in San Diego. Prof. Burbidge began his editorial with:

> "Big bang cosmology is probably as widely believed as has been any theory of the universe in the history of Western civilization. It rests, however, on many untested, and in some cases untestable, assumptions. Indeed, big bang cosmology has become a bandwagon of thought that reflects faith as much as objective truth."

He went on to say

> "Younger cosmologists are even more intolerant of departures from the big bang faith than their more senior colleagues are. Worst of all, astronomical textbooks no longer treat cosmology as an open subject. Instead the authors take the attitude that the correct theory has been found."

Burbidge then explained the basic reasons for the bandwagon mentality:

"Powerful mechanisms encourage this conformity. Scientific advances depend on the availability of funding, equipment, and journals in which to publish. Access to these resources is granted through a peer review process. Those of us who have been around long know that peer review and the refereeing of papers have become a form of censorship. It is extraordinarily difficult to get financial support or viewing time on a telescope unless one writes a proposal that follows the party line."

"A few years back, Halton C. Arp was denied telescope time at Mount Wilson and Palomar Observatories because his observing program had found, and continued to find, evidence contrary to standard cosmology."

"Unorthodox papers often are denied publication for years or are blocked by referees. The same attitude applies to academic positions. I would wager that no young researcher would be willing to jeopardize his or her scientific career by writing an essay such as this."

Eric Lerner and Nobel Laureate Hannes Alfven

In 1991, Eric Lerner wrote the book, *The Big Bang Never Happened* [16]. He was strongly supported in this book by Nobel laureate Hannes Alfven (1908-1995), who was the father of modern plasma physics. The stimulus for the book was the strong opposition that Hannes Alfven and other plasma physicists had received in their attempts to publish papers in astronomical journals that related plasma physics to cosmology.

There is abundant evidence that plasma physics effects have greatly influenced the development of our solar system, and have caused the rotation of galaxies and stars. Nevertheless, these papers have been continually rejected by the Big-Bang establishment, which controls astronomical literature and research funds.

Lerner [16] gives a detailed criticism of Big Bang research, which is only briefly summarized here. His book is highly recommended to show the serious problems in the field of astronomy today.

The Big Bang Age Dilemma

A serious weaknesses of the Big Bang theory is the universe age dilemma. According to the Big Bang theory, the age of the whole universe can be no greater than about 15 billion years, yet there is strong evidence that some stars in our galaxy are nearly 15 billion years old. An article in the May 2001 *Scientific American* described recent studies of this issue, and stated (page 53) that the age of the oldest stars is no less than 11.5 billion years and no greater than 14.5 billion years.

It has been claimed that some stars appear to be older than our universe. Although recent Big Bang studies have disputed this claim, there is little margin in the calculations. The development of stars and galaxies must have proceeded in an extremely efficient manner for our whole universe to have been formed within the 15 billion year maximum age of our universe.

The Big-Bang age dilemma is actually much worse than this discussion suggests. As explained by Eric Lerner [16] (pps 15-32), recent astronomical studies have shown that the universe is not at all uniform. In 1986, Brent Tulley, a University of Hawaii astronomer, (assisted by J. R. Fischer) found that almost all galaxies within 1.5 billion light-years are concentrated into huge ribbons, typically one billion light-years long, 300 million light-years wide, and 100 million light-years thick. These ribbons contain curling filaments, a few million light-years thick, extending for hundreds of millions of light-years.

This study evolved from a mapping of individual galaxies out to 100 million light-years. In making this map, it was assumed that the redshift of a galaxy specifies its distance. From this map Tulley and Fischer found (with about 20 exceptions) that all of the thousands of galaxies are concentrated into filaments, a few million light-years across. These filaments extend for hundreds of millions of light-years, beyond the limits of the map. While performing a later mapping study, astronomer Margaret Haynes concluded after examining the curling galactic filaments, *"The universe is just a bowl of spaghetti."*

To expand his study to 1.5 billion light-years, Tulley mapped clusters of galaxies, because there are millions of individual galaxies within that range, too numerous to be mapped individually. From this cluster map Tulley discovered his huge ribbons.

About 1990, Tulley's findings were confirmed by several astronomer teams. The most dramatic is by Margaret J. Geller and John P. Huchra of the Harvard Smithsonian Center for Astrophysics, who are mapping individual galaxies out to 600 million light-years, about 200 times as many as in the Tulley-Fischer map of individual galaxies. In their preliminary results, they displayed what they call the "Great Wall", a huge ribbon of galaxies stretching across the region mapped, a distance of 700 million light-years. This ribbon, which is 200 million light-years wide and 20 million light-years thick, corresponds closely to a ribbon mapped by Tulley using clusters of galaxies. Galaxy density inside the ribbon is 25 times greater than outside.

These results seriously contradict the Big Bang theory in two ways: (1) it would probably take at least 150 billion years to form these gigantic structures; and (2) the Big Bang theory predicts a very uniform universe, not the spaghetti-like and ribbon-like structures that are observed.

Lerner [16] (page 23) explains the universe age problem. Except for the general Hubble expansion of the universe, the maximum local velocity of all galaxies is less than 1000 km/sec, which is 1/300 of the speed of light. Since the time of the postulated Big Bang, a galaxy could move only a distance of 15 billion light-years divided by 300, which is 50 million light-years. However, if the universe were uniform after the Big Bang, galaxies would have had to move 270 million light-years to form the huge ribbons. The age discrepancy is worse than these numbers suggest, because time must be allowed for a galaxy to accelerate and decelerate.

Mythological Philosophy of Big Bang Research

The cosmic microwave radiation, which is used as primary evidence to support the Big Bang theory, is actually a serious liability to the theory, because it is far too smooth and uniform. It follows the theoretical blackbody spectrum to very high accuracy, and the radiation arrives with high uniformity from all directions. If this microwave radiation is the cooled relic of light radiated from very hot matter existing 300,000 years after the Big Bang, the universe must have been extremely uniform when the energy was radiated. How did the universe change from that extremely uniform state to the spaghetti-like and ribbon-like universe of today, during the short time following the Big Bang? The 15 billion year age of the universe claimed by Big Bang proponents is at least 10 times too short.

Let us ask the more realistic question: What process could cause galaxies to arrange themselves into these spaghetti-like and ribbon-like structures? To this question, Lerner [16] (pps. 39-49) gives a clear scientific answer. For many years, Hannes Alfven, a Swedish Nobel laureate and virtual founder of modern plasma physics, has proposed that electrical currents flowing through the ionized plasma of space produce strong magnetic forces that greatly affect the development of galaxies.

The thin gas of space is ionized, producing what is called plasma. This means that electrons are separated from the atoms, and so can move freely through the plasma to form electric currents. Although the electric current flowing through a square meter of area is very small, an enormous current can flow through the huge area associated with a typical star, which is several light-years wide. The electric currents flowing through the ionized plasma of space can generate magnetic fields of very high energy, which interact with the magnetic field of the star to alter the motion of the star. The electric and magnetic fields of the plasma currents tend to produce vortices that could cause the rotation and spiral shape of a galaxy. Over intergalactic distances, the electric and magnetic fields could create the spaghetti-like filaments into which galaxies are arranged.

Alfven has shown in laboratory experiments that instabilities cause plasma currents to form themselves into swirling electrical currents that twist relative to one another like the strands of a rope. The effects of such currents can be seen in the aurora, or "northern lights". They are also seen in a gaseous nebula, which is a mass of heated plasma surrounding a group of stars. On an intergalactic scale, plasma electric currents could produce the filament arrangement of galaxies, and the larger ribbon-like structures.

Much more information is given by Lerner [16] concerning the tremendous cosmological implications of plasma physics, which has been pioneered by Nobel laureate Hannes Alfven.

The Big Bang cosmologists have ignored or dismissed plasma theory, and few have even bothered to read about it. The well-known cosmologist, P. James E. Peebles (called the "father of modern cosmology" by *Scientific American*) stated that Alfven's ideas are *"just silly"*. His colleague at Princeton, Jeremiah Ostriker, commented, *"There is no observational evidence that I know of that indicates electric and magnetic forces are important on cosmological scales."*

Alfven, as well as lesser-known plasma physicists, have repeatedly had their papers rejected by astrophysical journals because they contradict Big Bang wisdom. Alfven commented, *"I think the Catholic Church was blamed too much for the case of Galileo — he was just a victim of peer review"*.

As evidence against the Big Bang has mounted, Big Bang cosmologists have shrugged it off, and have proceeded to devise more

10. Evidence Against the Big Bang

and more elaborate ad-hoc theories to bypass the evidence. Joseph Silk, who has written three books on the Big Bang, stated flatly

> "It is impossible that the big bang is wrong. Perhaps we'll have to make it more complicated to cover the observations, but it is hard to think of what observations could refute the theory itself."

Lerner [16] (page 54) responded to this with the following, which must also have reflected the convictions of Nobel laureate Alfven, who helped Lerner greatly in his book:

> "This attitude is not at all typical of the rest of science, or even of the rest of physics. In other branches of physics, the multiplication of unsupported entities to cover up a theory's failure would not be tolerated. The ability of a scientific theory to be refuted is the key criterion that distinguishes a science from metaphysics. If a theory cannot be refuted, if there are no observations that could disprove it, then nothing can prove it — it cannot predict anything; it is a worthless myth."

Lerner (p. 56) quotes the following words by Alfven on the myth issue:

> "The cosmology of today is based on the same mythological views as that of the medieval astronomers, not on the scientific traditions of Kepler and Galileo."

Big Bang theorists treat the Einstein gravitational field equation as their ultimate truth (their infallible "Bible"). As reported by Lerner (p. 163), cosmology theorist George Field stated his Big Bang philosophy as:

> "I believe the best method is to start with exact theories, like Einstein's, and derive results from them."

This philosophy of modern cosmology sharply contrasts with legitimate science, which demands consistency between theory and observation. It is the philosophy of mythology, not of science. Lerner (p. 162) states:

> "Entire careers in cosmology have now been built on theories that have never been subject to observational tests, or have failed such tests and have been retained nonetheless."

As Lerner (p. 127) points out, the mythology of modern cosmology is based on the "myth of Einstein". He explains Alfven's concepts with:

> "It is quite ironic that the triumph of science [from relativity theory] led to the resurgence of myth. The most unfortunate effect of the Einstein myth is the enshrinement of the belief, rejected for four hundred years, that science is incomprehensible, that only an initiated priesthood can fathom its mysteries."

Lerner quotes the following words of Alfven:

> "The people were told that the true nature of the physical world could not be understood except by Einstein and a few other geniuses who were able to think in four dimensions. Science was something to believe in, not something that should be understood. Soon the best sellers among the popular science books became those that presented scientific results as insults to common sense. One of the consequences was that the boundary between science and pseudo-science began to be erased. To most people, it was increasingly difficult to find any difference between science and science fiction."

Since the start of the wide acceptance of the Big Bang theory about 1970, cosmology theorists have been struggling with its many serious conflicts with observational evidence. Lerner [16] (pps 150-163) gives an excellent discussion of the increasingly complicated, bizarre, and arbitrary hypotheses that have been developed since 1965 to accommodate the more and more evident conflicts of the Big Bang theory with observed data.

The Modern Big Bang Singularity

Astronomers today endorse the modern Big Bang singularity concept. The primary evidence against the modern Big Bang theory is that the singularity concept drastically conflicts with physical evidence and was absolutely rejected by Einstein.

Chapter 11

Weaknesses of the Einstein Theory

Although the principles of the Einstein theory are sound, the Einstein gravitational field equation, which specifies that theory, has serious weaknesses. These were not apparent during Einstein's lifetime because of the great mathematical complexity of the Einstein equations. The Schwartzschild limit was an annoying problem, but Einstein dismissed it as physically irrelevant, since it requires a mass-to-radius ratio that is one-quarter million times greater than the maximum value experienced within our solar system.

In 1939, Oppenheimer and Snyder predicted that a star must collapse to form a *"black-hole"* singularity if its mass-to-radius ratio exceeds the Schwartzschild limit. Einstein flatly rejected this concept, because it drastically violates physical evidence. He believed that the Einstein gravitational field equation does not predict a singularity. No one could refute Einstein until computers became widely available in the 1960's, about a decade after his death.

The physically impossible Schwartzschild *black-hole* singularity indicates that the Einstein theory has a mathematical flaw. This chapter discusses several other weaknesses of the Einstein theory that have become apparent since Einstein's death. Probably the most serious flaw is that the Einstein gravitational field equation cannot achieve any more than a single-body solution.

Cannot Achieve a Two-Body Solution

Professor Carroll O. Alley

Professor Carroll O. Alley of the University of Maryland is one of the very few experts in General Relativity theory who has applied his knowledge to practical applications. Alley has supervised several experiments to test the validity of predictions derived from the Einstein theory. This has included laser measurements with retro-reflectors on the moon that have allowed several distances to the moon to be measured with laser beams to an accuracy of 3 centimeters. Another set of experiments measured the relativistic time delay in an atomic clock carried in an aircraft under several flight profiles. [Y10]

Alley is also intimately involved in applying General Relativity corrections to the Geophysical Positioning System (GPS). The GPS is an array of satellites operated by the U. S. Air Force to provide accurate position coordinates over the world for military and civilian navigation.

Alley became impressed with the Yilmaz theory and has cooperated with Yilmaz. Alley made an important contribution to this issue with his proof that the Einstein theory cannot achieve a two-body solution. Let us examine this astonishing claim.

The Single-Body Schwartzschild Solution

The Schwartzschild analysis of the Einstein theory was only a *single-body solution*. It merely considered the gravitational field of one body, a single star. When it was applied to calculate the relativistic advance of the Mercury orbit, the analysis assumed a test mass in the Mercury orbit that had absolutely no effect on the gravitational field.

The orbit of the planet Mercury advances by 1.39 arc seconds for every orbit. This means that the axis of the elliptical Mercury orbit rotates 1.39 arc seconds for every orbit of Mercury, and this rotation is in the direction of the Mercury motion. This advance of the Mercury orbit was determined by measuring the cumulative advance of the orbit over many years.

Of this 1.39 arc-second advance per orbit of Mercury, 1.29 arc seconds can be calculated using Newton's theory of gravity by considering the forces exerted on Mercury by other planets. The remaining 0.10 arc-second per orbit of the Mercury advance was calculated from the Schwartzschild solution. The 1.29 arc-second per orbit advance of the Mercury orbit caused by other planets could not be calculated from the Schwartzschild solution, because that solution could only consider the effect of the gravitational field of the sun.

As Mercury revolves around the sun, the gravitational field of Mercury causes the center of the sun to wobble slightly. The Schwartzschild analysis cannot include this effect, because it is only a *single body solution*. It can only account for the gravitational field of a single body, the sun. The gravitational field considered in the analysis is

11. Weaknesses of the Einstein Theory   125

not affected by the mass of any planet, including Mercury.

Einstein recognized that his analysis of the Mercury orbit used a *single-body solution*. He and other scientists were content with this limited analysis because it was not practical during Einstein's lifetime to achieve a multi-body solution with the very complicated equations of the Einstein theory. A multi-body solution of the Einstein equations would require a non-diagonal metric tensor, which would result in millions of terms in the analysis.

Einstein assumed that his theory could yield a multi-body solution, but the following discussion shows that it cannot. **More precisely, the Einstein theory cannot yield an interactive multi-body solution; it cannot allow two bodies to interact gravitationally.**

The Analysis of Professor Alley

Many computer studies using the Einstein theory have appeared to achieve multi-body solutions. However these studies employ artifices that help to make the iterative computer programs converge to solutions. These artifices are inserting results into the solutions that are not actually coming from the Einstein gravitational field equation.

Alley has avoided this problem by considering a simple physical model that can be solved analytically by the Einstein theory. He calculated the gravitational attraction between a pair of infinite slabs of matter separated by a fixed distance. He found that the Einstein theory predicts that there is no gravitational attraction between the two slabs.

This configuration is physically similar to models used in electronics to calculate capacitance. By assuming that the dimensions of the slabs are infinite relative to the separation between the slabs, one can ignore edge effects. This results in a simple theoretical model to which one can apply the Einstein theory analytically. [Y10, Y12]

As the Alley analysis shows, the Einstein theory predicts that there is absolutely no gravitational force between the slabs. This result conflicts with Newton's law of gravitational attraction and with experimental evidence. Chapter 4 described the experiment by Henry Cavendish, which determined the gravitational constant G by measuring the gravitational attraction between lead spheres. Alley's analysis directly conflicts with the Cavendish experiment.

Reason for Failure to Achieve a Two-Body Solution

In *Story* [4], Chapter 13 gives a physical explanation for why the

Einstein theory does not predict gravitational attraction between the two slabs in Alley's analysis. A serious limitation of the Einstein gravitational field equation is that it lacks a tensor to characterize the energy of the gravitational field. Without this tensor, the Einstein theory cannot predict interactive gravitational effects between two bodies.

The success of the Schwartzschild solution has disguised the fact that (1) it can only describe relativistic effects of the gravitational field for a single body, and (2) it gives reliable predictions only in weak gravitational fields. The Schwartzschild limit shows that the Einstein equations do not work in an intense gravitational field, and the Alley analysis shows that they do not allow an interactive multi-body solution.

Conservation of Matter-Plus-Energy

In order to yield realistic predictions, a relativistic theory must achieve conservation of matter-plus-energy. Since matter can be converted into energy, and vice-versa, it is the sum of matter-plus-energy that must be conserved. This conservation is achieved by placing appropriate constraints on the energy-momentum tensor.

This issue is discussed in Appendix E of *Story* [4]. Although the Einstein and the Yilmaz theories both have energy-momentum tensors, these tensors are not exactly the same. Yilmaz has proven that the energy momentum tensor for the Yilmaz theory always achieves conservation of matter-plus-energy, but the energy-momentum tensor for the Einstein theory generally does not, despite the common belief that it does.

Multiple Solutions from the Einstein Theory

A confusing aspect of the Einstein theory is that it can yield multiple solutions for the same physical model. Constraints that are somewhat arbitrary must be included in a General Relativity analysis to achieve an answer. This problem is widely recognized by those performing General Relativity studies.

In contrast, the Yilmaz theory has a definite solution and can yield only one answer for a particular physical model. The time-varying Yilmaz theory incorporates general constraints, which assure that this condition is always satisfied.

Scientists who are familiar with the arbitrary adjustable parameters of the Einstein theory may find it difficult to recognize that the Yilmaz theory does not have this property. A prediction made by the Yilmaz theory depends only on the characteristics of the physical model on which it is based. It is not affected by assumptions made by the individual who is applying the Yilmaz theory.

Variation of Speed of Light with Direction

For tests that have been performed within our solar system, the Einstein and Yilmaz theories yield essentially the same results. Consequently all tests that have verified the Einstein theory are consistent with the Yilmaz theory.

However, there is one very sensitive test, which has been partially implemented, that could distinguish between the Einstein and Yilmaz theories. Yilmaz predicts that the speed of light measured locally in a gravitational field is the same in all directions, but the Einstein theory predicts it is different.

As we saw in Chapter 8, Yilmaz proved that for the speed of light measured locally to be independent of direction, the product ($g_{00}g_{11}$) must be equal to -1, and the elements g_{11}, g_{22}, g_{33} must be equal. These conditions are satisfied by the Yilmaz theory but not by the Einstein theory.

It is very difficult to measure with sufficient accuracy the variation of the speed of light with direction, because the test cannot use two-way transmission of light. As explained by Prof. Carroll O. Alley [Y10], preliminary experiments were performed under his direction to perform this measurement. This involved the transportation of an atomic clock, back and forth between the U.S. Naval Observatory and the NASA Goddard Optical Research Facility, which are 21.5 km apart. The transported clock was used to synchronize atomic clocks in the two locations.

The initial data were very promising. However, government funding was cancelled before measurements could be made with the required accuracy to obtain definitive results. Why was the modest funding for this very important experiment cancelled?

This experiment was discussed by Ivars Peterson [30] in a 1994 *Science News* article: "A New Gravity: Challenging Einstein's general theory of relativity".

The Einstein Theory Is Not Rigorous

This discussion has shown that there is a flaw in the Einstein gravitational field equation, which characterizes the Einstein theory. Because of the great mathematical complexity of that equation, its weaknesses have been obscured.

The success of the Schwartzschild solution disguised the fact that it is only a single-body solution that gives reliable results only in a weak gravitational field. The Schwartzschild limit indicates that the Schwartzschild solution does not work in an intense gravitational field, even for a single-body solution. The Alley analysis proves that the Einstein equations cannot yield an interactive multi-body solution.

The Alley analysis proves that the Einstein gravitational field equation is not mathematically rigorous. Consequently, confusing results and nonphysical predictions have been derived from the Einstein theory. The physically impossible singularity predictions of Big Bang cosmology are examples of this.

The fact that the Einstein gravitational field equation is not rigorous should not diminish our great respect for the Einstein genius. He established a broad and solid foundation of principles for building his Relativity theory. The fact that he did not achieve his goal does not detract from the validity of that scientific foundation.

Yilmaz built on the relativistic principles of Einstein. Having a different perspective, he could see things that Einstein had missed, and thereby was able to derive a rigorous solution to the principles that Einstein had established. With this rigorous solution, Yilmaz proved that Einstein's relativistic principles were correct. The Yilmaz refinement of the Einstein theory eliminates the physically impossible singularity predictions that have been derived from the Einstein gravitational field equation.

The Yilmaz theory was published over 40 years ago, yet has been ignored by countless scientists who have focused their careers around the very complex equations of the Einstein theory. With the advent of powerful computers, the Einstein equations, with their enormous complexity, have provided the basis for limitless theoretical physical studies, particularly in cosmology. If it should become widely recognized that the Einstein gravitational field equation is not rigorous, this enormous body of theoretical research would be reduced to a mathematical exercise, devoid of physical meaning. Therefore scientists engaged in General Relativity research find it convenient to ignore the Yilmaz theory.

Finding the Truth

Many readers may lack the scientific background to assess the validity of claims that the Einstein equations are not rigorous.

11. Weaknesses of the Einstein Theory   129

Nevertheless, a general reader has a powerful tool that can determine where the truth lies. That tool is common sense.

Do you honestly believe that at the instant of the Big Bang our whole observable universe, containing matter equivalent to 20 billion times the mass of one trillion suns, could have begun as a body so compact it was microscopic in size?

Astronomers insist that they are finding black holes. However, the black hole concept requires that the star that lies within a black hole must be a singularity having an infinite density of matter.

Do you believe in the existence of a black hole, which contains a star that is squeezed until its diameter is zero and its density of matter is infinite?

We have discussed the neutron star, with its enormous density. Yet there is direct evidence that neutron stars actually exist. There is also extensive physical evidence that this same density, **one billion tons per teaspoon**, exists here on earth within the nucleus of every atom. Although neutron-star density may seem unbelievable, it is consistent with physical evidence.

A neutron star consists of tightly packed neutrons and has reached the greatest possible density of matter. There is no room left for further contraction. The vastly greater densities of matter that are predicted by modern cosmologists in their black hole and Big Bang singularities are drastically inconsistent with physical evidence, and were absolutely rejected by Einstein.

It should be obvious that modern Big Bang cosmology makes no sense. Regardless of the prestigious scientific credentials of the proponents of the modern Big Bang theory, their singularity concept is intellectually bankrupt. ***The common sense of the reader represents a wisdom that far exceeds the bandwagon mentality that now controls the field of astronomy.***

Understanding Our Universe

What does this tell us about the creation of our universe? It should be clear that the modern Big Bang theory is so far removed from physical reality it is not a viable explanation of cosmology.

The reason that cosmology is in such confusion is that the Einstein theory is being used as the theoretical foundation for cosmology models. However, we have seen that the Einstein equations are not rigorous and cannot be used to derive reliable cosmological predictions. The Yilmaz theory has refined the Einstein theory and has thereby achieved a rigorous theory of gravity. The Yilmaz theory can be used as the basis for a meaningful study of the universe.

As we reject the Einstein equations, we are not diminishing the genius of Albert Einstein. The revolutionary principles of Relativity that Einstein established are correct. The fact that Yilmaz has derived a rigorous theory of gravity from these principles has justified the validity of the Einstein principles.

The Yilmaz theory is a refinement of the Einstein theory that has achieved the goal that Einstein sought. This refined Einstein theory gives us a powerful mathematical tool for understanding physics, and with it a scientific basis for exploring the universe.

The Yilmaz cosmology model is a very simple application of the Yilmaz theory to cosmology. It assumes that the universe has a constant average density of matter that does not change with time. This simple model predicts that the universe should expand, just as Hubble observed. Even though gravity normally forces masses to attract one another, relativistic gravitational effects should make the universe expand. This concept is explained in Chapter 12.

Chapters 12 and 13 describe the predictions of the universe that are derived from the Yilmaz cosmology model. Are these predictions correct? Unlike the Big Bang theory, the Yilmaz cosmology model does not violate any physical law, and is based on a rigorous theory of gravity.

This cosmology theory provides a new and exciting picture of the universe. This picture rests on much firmer scientific ground than the singularity concept of the universe that is being portrayed as truth by modern Big Bang cosmologists.

Einstein's Search for a Unified Field Theory

After developing his General theory of Relativity, Einstein devoted most of his efforts in a search for a *Unified Field Theory*. Initially, this would have combined into a single integrated theory the principles of electromagnetic fields and gravitational fields. Later, as knowledge of the atomic nucleus evolved, he also attempted to include the effects of atomic nuclear fields.

Einstein realized that if he could solve this problem he would make a revolutionary advance in physics. He struggled with this task to his last days but never succeeded. A fundamental problem that Einstein faced in this quest is that his General theory of Relativity conflicts with quantum mechanics. However, the Yilmaz has proven that his refinement of the

Einstein theory is consistent with quantum mechanics.

Since the Yilmaz theory is a rigorous solution of the principles that Einstein established in deriving his General theory of Relativity, it has enormous potential. We have examined the cosmological implications of the Yilmaz theory, but this theory has far greater possibilities. Einstein had the wisdom to realize that a *Unified Field Theory* may be achievable, and this would lead to great advances in physics. Since the Yilmaz theory embodies the principles of the Einstein theory and is consistent with quantum mechanics, it may well provide the key for achieving Einstein's elusive *Unified Field Theory*.

Chapter 12

The Yilmaz Cosmology Model

Cosmological Implications of the Yilmaz Theory

Astronomy today is in a state of confusion, because cosmology studies are based on the conviction that the Einstein gravitational field equation is absolutely accurate, even though this equation yields physically impossible singularities. Einstein clearly stated that his General theory of Relativity would only hold approximately under conditions of extreme density of matter, and so cannot be used to predict a physical singularity.

The Yilmaz theory wipes away this cloud of confusion. The Yilmaz theory is a refinement of the Einstein theory, which provides a rigorous solution to Relativity principles. With the Yilmaz theory, the physically impossible singularity predictions are eliminated and a reasonable picture of the universe emerges.

The Yilmaz theory is applied to cosmology by making a simple postulate. The universe is assumed to have a constant average density of matter that extends to infinity and does not change with time. Although the universe is lumpy, the assumption of uniform mass density should give a good first approximation for the relativistic effects produced by gravity. The resultant Yilmaz cosmology model yields an exciting new picture of our universe.

In agreement with the Steady State Universe theory, the Yilmaz cosmology model predicts that our universe is infinitely old, and that matter is continuously created to compensate for the Hubble expansion. However, unlike the Steady-State Universe theory, the Yilmaz cosmology model predicts a definite source for the newly created matter. The matter is derived from energy radiated from stars. Conservation of matter and energy is achieved within an enormous universe that has a constant size.

12. The Yilmaz Cosmology Model

Big Bang cosmologists have loudly claimed that cosmic microwave radiation proves that a Big Bang must have occurred. However, unlike the Steady-State Universe theory, the Yilmaz cosmology model also predicts this cosmic microwave radiation. As shown in Appendix B, it does so more accurately than the Big Bang theory.

The Yilmaz cosmology model predicts that the universe must expand. The Hubble expansion is a natural relativistic effect that is directly caused by gravity. How can gravity, which always causes masses to attract one another, force the universe to expand? This chapter gives a simple physical explanation for this astounding effect.

One may ask, "Is the Yilmaz cosmology model correct?" There are strong reasons to take this cosmology model seriously. For many years astronomers have been claiming with complete confidence that they understand how our universe was created, but their explanations include singularities that can best be described as science fiction. There is certainly much more reason to believe the Yilmaz cosmology model than these mathematical fantasies.

When the Einstein theory is used to analyze a particular physical model, different answers can result depending on the assumptions made in implementing the analysis. In contrast, the Yilmaz gravitational theory gives unique answers, which depend only on the characteristics of the physical model being studied. The Yilmaz gravitational theory does not allow the arbitrary adjustable parameters used in General Relativity calculations. The fact that the Yilmaz theory predictions are unique is another reason to take the Yilmaz cosmology model seriously.

Yilmaz presented his cosmology model in the first paper on his theory presented in 1958, but has not pursued the model further. The author has extended this cosmology model. The analyses were presented in Appendices C and D of *Believe* [1]. This chapter summarizes the material of Appendix C, which describes the basic characteristics of the Yilmaz cosmology model. The material of Appendix D describes the cosmic microwave radiation predicted by the model, and is summarized in Appendix B of this book. This material in Appendices C and D of *Believe* [1] is also given in the *Website* [3], Page 4.

The material in *Believe* [1] assumed a Hubble constant of 25 km/sec per million light-years. The plots of this book have been modified for a Hubble constant of 20 km/sec per million light-years. Let us now examine the predictions derived from the Yilmaz cosmology model.

Description of the Yilmaz Cosmology Model

Reduction of Speed of Light, Clock Rate, and Spatial Dimensions

We saw in Figs 9-1 and 9-2 of Chapter 9 that the gravitational field of a star causes the speed of light to decrease, it causes a spatial dimension to contract, and it causes a clock to run slower. The Yilmaz Cosmology Model predicts that the gravitational field produced by matter in the universe should cause similar relativistic effects as we look out into space. The results are shown in Fig. 12-1. The solid curve shows how the speed of light should decrease with distance, and the dashed curve shows how a clock rate and a spatial dimension should decrease with distance.

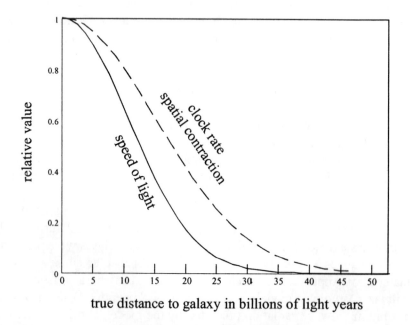

Figure 12-1: Apparent speed of light (solid), clock rate (dashed), and spatial contraction (dashed), versus distance to a galaxy

The horizontal scale of Fig. 12-1 shows the true distance to a galaxy expressed in billions of light years. The Big Bang theory assumes that our observable universe extends only to 15 billion light years; beyond that limit galaxies are theoretically receding faster than the speed of light and so cannot be seen. In contrast, the Yilmaz cosmology model predicts

12. The Yilmaz Cosmology Model   135

that galaxies can be observed at true distances much greater that 15 billion light-years.

In Fig 12-1, the apparent speed of light drops to half its normal value for a galaxy at a true distance of 13 billion light years. Apparent spatial dimensions and apparent clock rated drop to half of their normal values for a galaxy at a true distance of 18 billion light years. These apparent values of speed of light, spatial dimension, and clock rate are the values that we observe from earth. An observer located at the distant galaxy would experience the proper (or "true") values.

These relativistic effects are similar to those caused by velocity in the Special Relativity discussion of Chapter 6. The earth observer experiences the apparent dimension and apparent clock rate of the measurement equipment on the space ship. However, the space ship observer experiences the proper (or "true") dimension and clock rate for his equipment.

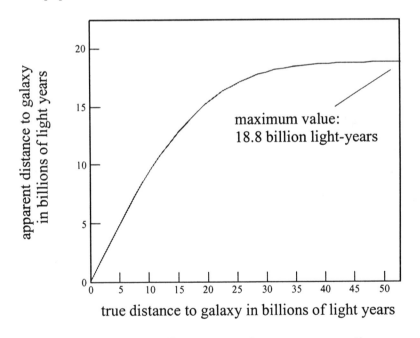

Figure 12-2: Apparent distance to galaxy verses true distance

The plot of spatial contraction in Fig 12-1 shows that a dimension appears to contract with distance. Consequently, the apparent distance to a galaxy is less than the true distance. Figure 12-2 gives a plot of the apparent distance to a galaxy versus the true distance. Because of the strong contraction of spatial dimensions at great distances, the maximum apparent distance to any galaxy is finite, even though the model assumes that true galaxy distance extends to infinity. The maximum apparent distance to a galaxy is

$$\text{Maximum apparent galaxy distance} = \sqrt{[\pi/2]}\ r_0$$

The parameter r_0 is 15 billion light-years, and so this maximum apparent distance is 18.80 billion light-years.

The parameter r_0 is defined as (c/H_0), where H_0 is the Hubble constant. For our assumed Hubble constant (20 km/sec per million light years), r_0 is equal to 15 billion light years. For the Big Bang theory, r_0 represents the radius of the observable universe, but it does not have this meaning in the Yilmaz cosmology model. In the Big Bang theory, we calculate the radius r_0 of the observable universe by assuming that the universe expansion rate is constant. With this assumption, r_0 is the distance at which the velocity of a galaxy would reach the speed of light. However, the Yilmaz cosmology model predicts that the universe expansion rate decreases with distance, and so the velocity of a galaxy never reaches the speed of light.

Figure 12-2 shows that the apparent distance of a galaxy cannot exceed a maximum value of 18.8 billion light years, even though the true distance to a galaxy can theoretically extend to infinity. No galaxy can appear to be more distant than 18.8 billion light years, relative to measurements made from earth.

For a galaxy that is close to the apparent limit of 18.8 billion light-years, there is strong contraction of the apparent size of the galaxy. Consequently the apparent density of matter becomes very high as the limit at 18.8 billion light-years is approached. This effect is shown in Fig. 12-3.

Figure 12-3 shows how the density of the universe appears to vary with apparent distance from the earth. The darker the picture, the higher is the apparent mass density. The circles are contours of constant apparent mass density, and correspond to relative density values of 1.5, 3, 10, 30, 100, and 1000. The apparent distance of the inner circle is 7.5 billion light-years, and that of the periphery is 18.8 billion light-years.

Figure 12-3 shows how the universe appears as seen by an observer located on earth. Within 7.5 billion light years, the universe appears to be highly regular. However, at greater distances relativistic gravitational effects strongly compress the space. Because of this compression, the apparent density of matter becomes extremely high as the apparent distance approaches 18.8 billion light years.

12. The Yilmaz Cosmology Model 137

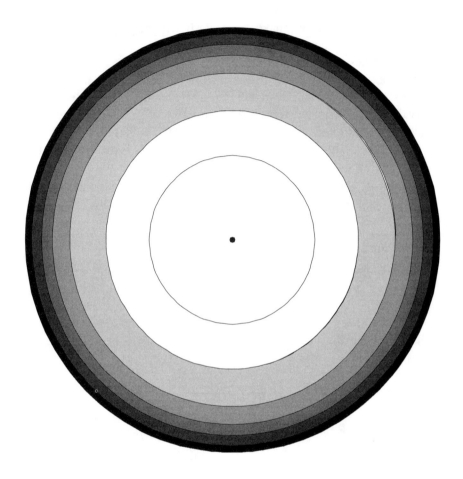

Figure 12-3 Apparent relative mass density of universe seen from earth; boundaries at density values of 1.5, 3, 10, 30, 100, and 1000; minimum and maximum radii at 7.5 and 18.8 billion light years

We should realize that the "apparent" effects shown in Figs. 12-1 to 12-3 are real effects. As was explained in our discussion of Special Relativity, *Reality is Relative*. There is no such thing as *absolute reality*. Any observation that we make from earth will display the "apparent" characteristics of the universe that we have examined.

Nevertheless we can still consider a "true" picture of the universe where galaxies are at a "true" distance from the earth. This "true" picture gives us a simple absolute model for understanding the universe. An interval of "true distance" represents an interval of "proper distance" that would be measured in proper coordinates situated at that location.

Figure 12-2 shows that a distant galaxy appears to be closer than it actually is. Nevertheless, the apparent time for the light to reach us from the galaxy is not reduced, because the apparent speed of light and the apparent clock rate also decrease with distance. Combining the spatial contraction, speed of light, and clock rate plotted in Fig. 12-1 shows that the apparent time for light to travel between two distant points is the same as if there were no relativistic effects.

For example, consider a distance where the spatial contraction is ½. The relative speed of light is ¼ and the clock rate is ½. With distance reduced to ½, light should take twice as long to travel between two points if the speed of light is ¼. However, the apparent clock rate is cut in half, and so the apparent time for light to travel between the two points is the same as if there were no relativistic effects.

The Hubble Expansion of the Universe

In General Relativity theory, the *geodesic equations* are used to calculate the trajectory of a planet or any other body, or of a particle, or even of a light photon. To verify the Einstein theory, the geodesic equations were applied to the Schwartzschild solution to calculate the bending of a light ray when it passes close to the sun, and to calculate the relativistic advance of the orbit of Mercury.

When the geodesic equations are applied to the Yilmaz cosmology model, they show that the universe must expand just as Hubble observed. A galaxy must recede at a velocity approximately proportional to distance. The results of this analysis are shown in Fig. 12-4. The solid curve is the ratio of apparent receding velocity of a galaxy relative to the apparent speed of light.

The dashed line in Fig. 12-4 shows for comparison the ideal Hubble law, in which galaxy velocity is exactly proportional to distance. The Hubble law is assumed by the Big Bang theory. According to the Hubble law, the velocity of a galaxy would reach the speed of light at a distance r_0 of 15 billion light-years, and would exceed the speed of light at greater distances.

For the Yilmaz cosmology model, the solid curve shows that the receding velocity of a galaxy approximates the ideal Hubble law plot out to about 5 billion light-years. Hence the universe should expand approximately according to the Hubble law within 5 billion light-years. At much greater distances, the apparent galaxy velocity gradually approaches the apparent speed of light, but never exactly reaches it.

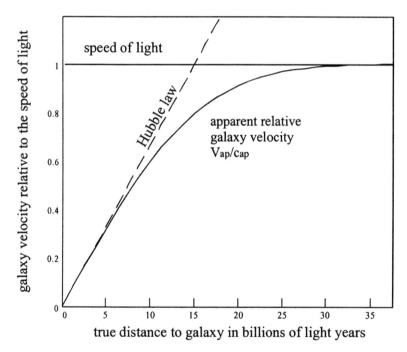

Figure 12-4: Apparent galaxy velocity relative to apparent speed of light, compared with Hubble law.

The solid curve in Fig. 12-4 represents the ratio V_{ap}/c_{ap}, which is the apparent receding velocity (V_{ap}) divided by the apparent speed of light (c_{ap}). It is this apparent velocity ratio that astronomers calculate from the Doppler wavelength shift of galaxy spectra.

To see how the universe is actually expanding, we need the ratio V/c, which is the true (or "proper") galaxy velocity V divided by the speed of light c measured on earth. The spatial contraction plot in Fig.

12-1 is equal to the velocity ratio V_{ap}/V, and Fig. 12-1 also gives the relative speed of light c_{ap}/c. The following plots are combined to obtain the V/c ratio: the plot of V_{ap}/c_{ap} in Fig. 12-4, the speed of light plot of c_{ap}/c in Fig. 12-1, and the spatial contraction plot in Fig. 12-1, which is equal to V_{ap}/V. The resultant plot of V/c is shown in Fig. 12-5.

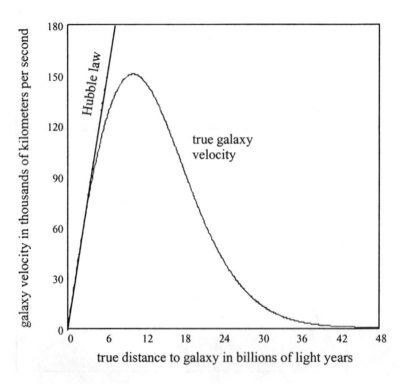

Figure 12-5: True galaxy velocity expressed in thousands of kilometers per second compared with Hubble law.

In Fig. 12-5, the speed of light c is set equal to 300,000 km/sec to give the true receding velocity V of a galaxy in absolute terms. This shows that the true velocity V of a galaxy reaches a maximum value of 150,000 km/sec at a true distance r of 12.5 billion light-years. This maximum value is half the speed of light c measured on earth.

At greater distances, the true galaxy velocity decreases, and becomes very small at large true distances. ***Thus Fig. 12-5 shows that over very large distances the universe does not expand. It demonstrates that the Hubble expansion is a local relativistic distortion of space, not a general expansion of the universe.***

12. The Yilmaz Cosmology Model   141

Figures 12-4 and 12-5 present some remarkable predictions concerning the universe that have been derived from the Yilmaz cosmology model. Figure 12-4 shows that the Hubble expansion of the universe is a natural relativistic effect that is directly caused by gravity. We do not need any exotic postulates to explain the Hubble expansion. This concept was discovered by Yilmaz in the mid 1950's when he first developed his gravitational theory. He was surprised to discover that his theory predicted the Hubble expansion.

Figure 12-5 shows an even more amazing effect, which the author discovered when he extended the analysis of the Yilmaz cosmology model. This shows that the Hubble expansion is a local effect. Over very large distances the universe does not expand. In other words, the universe expands locally about every point in the universe, but the universe does not get any bigger.

This is a revolutionary cosmological principle. It is not a postulate. It is a prediction of the Yilmaz cosmology model that evolved directly from its equations, which are based on the rigorous Yilmaz theory of gravity. The author was surprised when this principle was discovered.

How Can Gravity Make the Universe Expand?

The expansion of the universe displayed in Figs. 12-4 and 12-5 is a consequence of the Yilmaz gravitational theory. The analysis indicates that the Hubble expansion of the universe is caused by gravity. "How can this be?", you ask, "How can gravity, which always causes masses to attract one another, force the universe to expand?"

Part of the answer is that the Hubble expansion is a local effect. Over very large distances the universe does not expand. Yet we are still faced with the question, "How can gravity make the universe expand locally?"

The rigorous mathematical answer is that the geodesic equations, which characterize the effect of gravity, show that the universe must expand. This issue in discussed in Appendix C of *Believe* [1] and is proven mathematically in the *Website* [3], *Addendum* page 5, Chapter 4. Nevertheless we still would like an intuitive answer that makes sense physically.

Let us consider the following intuitive explanation. As was shown in Fig. 12-3, the whole universe appears to be compressed within a sphere having a radius of 18.8 billion light-years. The figure shows that the apparent density of matter is extremely high near the periphery of this sphere. We can assume that this outer shell of high-density matter is exerting gravitational force on matter that is inside the sphere. This gravitational force could be pulling matter toward the periphery, thereby producing the Hubble expansion. This gravitational attraction could cause a local expansion of the universe.

However, this simple intuitive explanation is not consistent with gravitational force as described by Newton. Suppose that the universe is modeled as a thin spherical shell, in which matter is evenly distributed over the shell. If Newton's law of gravity is applied to a mass element placed inside the shell, the gravitational forces on the mass are exactly cancelled. There is no net force attracting the mass element toward the spherical shell.

On the other hand, Newton's laws are only approximated in this relativistic model. Gravitational forces inside the shell do not cancel exactly, and so there can be a net gravitational force attracting matter toward the mass of the shell. This interpretation appears to explain how the attractive effect of gravity can cause the universe to expand locally.

This gives a simple intuitive explanation of how gravity could make the universe expand. On the other hand, as was stated earlier, the rigorous answer is that the universe expands locally because the geodesic equations show that it must.

Creation of Matter

The Yilmaz cosmology model assumes that the average density of matter does not change with time. In order to satisfy this requirement as the universe expands locally, the cosmology model requires that matter must be created to compensate for the expansion. Since matter and energy are equivalent, the creation of matter can be achieved by converting energy into matter.

The Yilmaz gravitational theory requires that matter-plus-energy must be conserved. Consequently, the Yilmaz cosmology model implies that energy is radiated across the universe to create matter that compensates for the Hubble expansion. This energy would be derived from matter in other parts of the universe.

As shown in Appendix A, the Yilmaz cosmology model predicts that one hydrogen atom is created every year within one cubic kilometer. The rate of conversion of energy into matter to achieve this creation of matter is equal to 5 microwatts of power continually converted into matter within a volume the size of the earth.

Cosmic Microwave Background Radiation

Proponents of the Big Bang theory strongly acclaim the discovery of cosmic microwave background radiation as a milestone in validating the theory. In 1965 this radiation was first detected in a sensitive communication antenna at Bell Laboratories. Much more accurate measurements were obtained from the Cosmic Background Explorer (COBE) satellite in 1989. These experiments show that microwave background radiation is emanating uniformly from all directions, and has the spectrum and intensity that would be emitted from an ideal blackbody at a temperature of 2.73 degrees Kelvin.

Cosmic microwave background radiation was predicted by Big Bang theorists, who claimed it to be the cooled relic of radiation emitted from hot plasma 300,000 years after the Big Bang. However estimates of the blackbody temperature by Big Bang theorists varied from 5 °K to 30 °K, and so the cosmic radiation was only predicted in a qualitative sense.

As shown in Appendix B, the Yilmaz cosmology model also predicts cosmic microwave radiation. Figure 12-4 indicates that at very large true distances the apparent galaxy velocity is very close to the apparent speed of light, and so the light radiated from distant galaxies should be Doppler shifted to very low frequencies.

The cosmic microwave radiation predicted by the Yilmaz cosmology model is equivalent (in both spectrum and intensity) to the emission from an ideal blackbody at a temperature between 2.1 °K and 3.4 °K. This range is highly consistent with the 2.73 °K blackbody temperature measured by the COBE satellite.

Uniqueness of Yilmaz Theory Predictions

The Yilmaz gravitational theory gives a rigorous and unique relativistic specification of the effects of gravity. The Yilmaz cosmology model applies this gravitational theory to a simple physical model of the universe. Since the Yilmaz theory is rigorous, and its predictions are unique, the description of the universe that has resulted from the Yilmaz cosmology model should be taken seriously.

For scientists accustomed to the Einstein theory, these principles may be difficult to understand, because the Einstein theory yields multiple, contradictory solutions. Since the predictions of the Yilmaz theory are unique, they are not affected by arbitrary assumptions made by the individual who is applying the Yilmaz theory.

Chapter 13

A Believable Picture of the Universe

Chapter 12 described the Yilmaz cosmology model, which was derived from the Yilmaz gravitational theory. Since the Yilmaz theory is a rigorous refinement of the Einstein theory, and its predictions are unique, we should take the Yilmaz cosmology model seriously. Let us examine the picture of the universe that is derived from this model.

The Implications of the Yilmaz Cosmology Model

The Yilmaz cosmology model accepts the basic principles of the Steady-State Universe theory, which assumes that the universe is infinitely old. The Steady-State Universe theory postulates that the universe has always appeared about as we see it today, and looks roughly the same from every point of the universe. As the universe expands, diffuse matter is created throughout space to compensate for the expansion. This matter is probably created one atom at a time as a hydrogen atom with a single proton and electron. The diffuse matter forms new stars and galaxies, and so the universe continually changes in detail, although its general appearance is eternal.

The Yilmaz cosmology model differs from the Steady-State Universe theory in that it explains the source of the created matter. It predicts that matter is derived from energy radiated from stars. How is this conversion of energy into matter achieved? Although we do not know this answer, this postulate does not violate any law of physics and is consistent with observational evidence.

The required rate of creation of matter is far too small to be observed directly. This rate is equivalent to one hydrogen atom created every year within a cubic kilometer. One hydrogen atom would be created every million years within a cube that is 10 meters on a side, the volume of a medium-size house. The rate of energy utilization required to create this matter is equivalent to the continual conversion of 5 microwatts of power into matter within a volume the size of the earth.

Because of gravitational attraction, the diffuse matter created throughout the universe continually congregates into enormous clouds of hydrogen. A cloud of hydrogen forms a galaxy, and bits of the cloud form individual stars.

Matter Derived from Gravitational Waves

It is postulated that energy radiated from stars in the form of electromagnetic waves (light, heat, X-rays, etc.) is converted into diffuse matter in space. This matter coalesces to form new stars and galaxies, and the process continues indefinitely.

However, this postulate alone could not result in a steady-state universe that lasts indefinitely. The universe would gradually deteriorate into black dwarf and neutron stars, and so would eventually die. There must be a process that converts the mass of these dead stars into energy.

The Einstein and Yilmaz theories both predict gravitational waves. As explained in *Scientific American*, April 2002 (pp. 62-71), scientists are performing elaborate experiments that attempt to measure gravitational waves. We postulate that the mass of a black dwarf or a neutron star is converted into gravitational waves that are radiated. By this means, the matter within these dead stars would gradually dissipate as radiated energy. The extreme densities of black dwarf and neutron stars might accelerate the generation of gravitational waves.

Therefore we postulate that energy is radiated from stars in the form of electromagnetic waves and gravitational waves, and this energy is converted into diffuse matter throughout the universe. During the lifetime of a star, less than 1 percent of its mass is converted into electromagnetic energy. Consequently our cosmology theory predicts that nearly all of the energy radiated from stars to form new diffuse matter comes from gravitational waves that are radiated from dead black dwarf and neutron stars.

We do not know how the energy radiated from stars might be transformed into matter. Nevertheless, this postulate is consistent with our laws of physics. In contrast, the Big Bang theory is based on the physically impossible postulate that all of the matter and energy of the universe was created *out of nothing*, at the instant of the Big Bang.

Our cosmology model envisions a universe of infinite age. New stars are steadily being formed, as energy radiated from older stars is converted into diffuse matter, which condenses to form new stars and galaxies. Because of the continual creation of matter, the universe is always changing, and so our universe stays eternally young even though it is infinitely old.

The Local Expansion of the Universe

An important aspect of the Yilmaz cosmology model is that the expansion of the universe is a local effect. Over very large distances the universe does not expand. The Yilmaz cosmology model predicts that the universe expands locally about every point, yet the over all size of the universe remains constant. **How is it possible for the universe to expand everywhere yet not grow any bigger?**

This relativistic concept evolved rigorously from the mathematics of the Yilmaz gravitational theory. It is not a product of speculation, and it surprised the author when it was discovered. Although this concept may conflict with our physical notions, it is no more counter-intuitive than the principles of Special Relativity explained in Chapter 6. The concept that two observers moving at different velocity measure the same speed of light, strongly conflicts with intuition.

A fundamental problem with the original Steady-State Universe theory is that its postulated universe continually grows larger, and has been doing so for an infinite period of time. In contrast, the Yilmaz cosmology model predicts a steady-state universe that has a constant size. Within this universe, matter-plus-energy is conserved.

The Second Law of Thermodynamics

Another question concerning the Yilmaz cosmology model is that its prediction of an infinitely old universe seems to conflict with the principles of thermodynamics. Scientists often use the *Second Law of Thermodynamics* to prove that our universe must eventually run out of available energy and so will die. This law of physics seems to refute the possibility that the universe can have an infinite age.

The United States Patent Office does not prohibit a patent on a perpetual motion machine. It merely requires that the inventor submit a working model of the invention before it can be patented. There are two types of perpetual motion machines, depending on which law of thermodynamics is violated. These two laws of thermodynamics can be expressed in simple terms as follows:

First Law: Energy can neither be created nor destroyed. Since matter and energy can be converted into one another, this is expressed in a more general form as: The sum of matter-plus-energy can neither be created nor destroyed.

Second Law: The availability of energy must continually decrease.

The *Second Law* is rigorously expressed in terms of a concept called *entropy*, which roughly means the *degree of disorder*. The *Second Law* states that *entropy* (the degree of disorder) must continually increase. However, the above definition gives a simpler interpretation of the *Second Law*.

The *Second Law* can be explained as follows. Consider a train locomotive driven by a steam engine. Water in the boiler is heated to generate steam, which pushes against the pistons to drive the wheels of the locomotive. It can be shown that the energy that can be extracted from the heated steam depends on the difference in temperature between the steam in the boiler and the temperature (100 °C) at which the exhausted steam condenses to water.

The gasoline engine in an automobile achieves much greater efficiency. Its energy output depends on the difference of temperature between the exploding gasoline in the cylinders and the temperature of the air into which the gasses are exhausted. Gasoline engines are run as hot as practical in order to increase their efficiency. Diesel engines achieve greater efficiency than regular gasoline engines because they operate at higher temperatures.

A motor can theoretically run from any source of heat, provided that there is a sink at a lower temperature into which the heat can be discharged. Useful energy cannot be derived from a heat source without a temperature differential.

With the passing of time our world becomes more uniform. Temperature differentials decrease, and so energy becomes less available. Our earth is replenished by energy radiated from our sun, which is generated by nuclear fusion occurring within the sun. Eventually the sun will run out of nuclear fuel, and will fade to become a black dwarf. After that occurs, the available energy on the earth will decay to zero.

Thus, in accordance with the *Second Law of Thermodynamics*, the availability of energy within our universe seems to be continually decreasing. Although matter-plus-energy remains constant within our universe, the availability of energy should theoretically decrease until the universe dies. How can we justify a universe that is infinitely old?

The Big Bang theory postulates that an enormous amount of energy and matter was created suddenly *out of nothing* at the instant of the Big Bang. This theory violates both laws of thermodynamics at the moment of creation.

The original Steady-State Universe theory postulates that matter is being created very slowly *out of nothing* to form diffuse matter in space. Like the Big Bang theory, this theory violates both laws of thermodynamics, but does so continuously in infinitesimal amounts.

The Yilmaz cosmology model requires the conservation of matter-plus-energy, and so it satisfies the First Law of Thermodynamics. But what about the Second Law?

Our model postulates that energy radiated from stars forms diffuse matter in space, which coalesces to create new stars. These new stars radiate energy, which forms new diffuse matter, and the process continues indefinitely. Matter-plus-energy is being conserved, as the *First Law of Thermodynamics* requires. However, the availability of energy does not decrease with time, and so the *Second Law of Thermodynamics* is violated. How do we justify this?

This contradiction is explained by the principle of *Relativity*, which allows the universe as a whole to violate the *Second Law of Thermodynamics*. Energy is radiated from distant regions of the universe to generate the matter that compensates for the Hubble expansion. Since *Reality is Relative*, relativistic effects can allow the *Second Law of Thermodynamics* to be violated by the universe as a whole, even though this law is satisfied locally.

Chapter 3 discussed the cosmological theory of Paul Marmet, which postulates that the Hubble redshift is an apparent effect, caused by photon collisions with hydrogen atoms. Marmet postulates that the Hubble redshift does not represent an actual expansion of the universe.

The Marmet cosmology theory seems doubtful because it requires a density of matter that far exceeds astronomical evidence. However, a more basic limitation of the Marmet theory is that it does not account for the Second Law of Thermodynamics. If the Marmet cosmology theory were correct, the Second Law would eventually cause the universe to run out of available energy and die. If the universe must eventually die, it must have had a beginning. What was that beginning?

Since the Hubble expansion was discovered in 1929, it has been regarded as an enigma that requires an explanation. **This book shows that the Hubble expansion is not an enigma. It is an essential feature if our universe is to endure forever.** The Hubble expansion supports the generation of new diffuse matter, which allows the universe to change continuously and thereby to stay eternally young although it is infinitely old. This new matter is derived from energy radiated from distant stars. The Hubble expansion is the basic mechanism for recycling matter in our universe, which converts matter from old dead stars into new material that forms new stars and galaxies.

In this process, relativistic effects compensate for the *Second Law of Thermodynamics* to keep the universe from running out of available energy.

Contents of the Universe

The "Observable" Yilmaz Universe

How large is our universe, and how much matter does it contain? As shown in Appendix B, the author's analysis indicates that cosmic microwave radiation is the Doppler-shifted effect of optical radiation emitted from galaxies at a true distance of 56 billion light-years. Since the predicted blackbody temperature agrees closely with the measured COBE value (2.73 °K), we can conclude that the universe should extend uniformly to a true distance of at least 56 billion light-years.

Light radiated from beyond 56 billion light-years cannot reach us, because it is absorbed by diffuse matter in space. Therefore we can regard 56 billion years to be the radius of the observable universe according to the Yilmaz cosmology model. Galaxies at this 56 billion light-year limit of our universe cannot be observed individually. The radiation from these galaxies is smeared to form the cosmic microwave radiation.

Appendix A computes the total matter within the observable Big Bang universe, which has a radius of 15 billion light-years. Since the observable universe for the Yilmaz cosmology model has a radius of 56 billion light-years, the ratio of the volumes of these two observable universes is $(56/15)^3$, which is a factor of 52. *This shows that the Yilmaz observable universe should have 52 times as much mass as the observable Big Bang universe.*

Appendix A calculates the following for the total mass of the observable Big Bang universe (with a radius of 15 billion light years):

from astronomical data: 19×10^{21} suns   (measured)
from Yilmaz cosmology model: 48×10^{21} suns   (theoretical)

The theoretical value is 2.5 times the measured value. With the large errors associated with these calculations, this represents very good agreement between theoretical and measured values of universe mass. *In words, the theoretical value represents a mass of 48 billion times one trillion suns within the 15 billion-light year radius of the Big Bang observable universe.*

To obtain the mass within the observable Yilmaz universe, the mass within the Big Bang universe should by multiplied by 52. This gives a total theoretical mass within the observable Yilmaz universe of 25×10^{23} suns. *In words, the theoretical mass within the observable Yilmaz universe represents a mass of 2.5 trillion times one trillion suns. The value derived from astronomical data is 40 percent of this value, or one trillion times the mass of one trillion suns.*

Is the Universe Infinite?

We may never be able to determine whether the size of the universe is finite or infinite. However it seems more comforting to believe that it is finite. We might assume that the universe folds back onto itself, and so it is finite even though it has no boundary.

At the limit of the Yilmaz observable universe, the apparent receding velocity of a galaxy is very close to the apparent speed of light. However, the apparent speed of light at that distance (56 billion light-years) is only 234 meters per second. The true (or proper) galaxy velocity is 280 km/sec, which is about 1000 times greater.

At a true distance of 75 billion light-years, which is beyond the observable limit of 56 billion light-years, the true receding velocity of a galaxy should be a mere 1 km/sec, which is only 1/30 of the velocity of the earth around the sun.

Let us assume that the universe folds back onto itself and has an effective radius of about 75 billion light-years. This means that if one could travel 75 billion light-years in any direction, one would reach the same point. With this postulate, the volume of the total universe would be $(75/56)^3$ times the volume of the observable Yilmaz universe, which is somewhat greater than a factor of 2. Therefore we double our estimate for the observable Yilmaz universe to obtain the contents of the total Yilmaz universe.

We estimate that our total universe has 5 trillion times the mass of one trillion suns. (One trillion is one million times one million.) Yet our sun has one million times the volume of our earth, which is so limitless to our senses that most people thought the earth was flat until the days of Columbus. From the point of view of a mere mortal, our universe might just as well be infinite.

Religious and Philosophical Implications of Our Picture of the Universe

The Biblical Story of Creation

The Big Bang instant of creation has often been related to the story of Creation in the Bible. Many have claimed that the Big Bang theory strongly reinforces this Biblical story. Nevertheless, we can find a solid basis for supporting the Biblical account of Creation without assuming that our whole universe began with a Big Bang 15 billion years ago.

Our sun was created 5 billion years ago. The earth, which defines our world, was created as a molten mass 4.6 billion years ago, and a cooled earth was formed 4.3 billion years ago. Our oceans were created from water supplied by meteorites. The photosynthesis from life containing chlorophyll created the oxygen in our atmosphere, which allows us to breathe. The scientific story of the creation of our sun and our earth, and the life on earth, is all that is needed to support the principles of the Biblical story of Creation.

This book is not claiming that the story of Creation in the Bible is an accurate account of creation. On the other hand, it is difficult to understand why the Biblical description of Creation has been interpreted as support for the Big Bang theory. The Bible begins with:

> *"In the beginning, God created the heavens and the earth. The earth was without form and void, and darkness was upon the face of the deep"* . . . *"And God said, 'Let there be light', and there was light."*

This is followed by a description of the development of our earth. Since the sun was created before the earth, the above quotation must be referring to the formation of our sun if the account is to have validity.

One could interpret this quotation as an explanation of the process that created our sun with its solar system. A cloud of hydrogen condensed until nuclear fusion was ignited and our sun began to shine. *"And God said, 'Let there be light', and there was light."* Initially, *"The earth was without form and void, and darkness was upon the face of the deep"*, a collection of dust and gas particles that coalesced to produce our earth.

Thus the Creation story of the Bible is consistent with the formation, 5 billion years ago, of our sun and its solar system, including our earth. Whether or not this creation of our sun and earth was preceded by the Big Bang creation of our whole universe, 10 billion years earlier, is not even hinted at in the Bible. **The Biblical story of Creation neither agrees nor disagrees with the Big Bang theory.**

Our Picture of the Universe

The Yilmaz cosmology model predicts that the universe as a whole has always existed, but individual stars and galaxies are continually being created. Matter is created to offset the local expansion of the universe. Since this matter is derived from energy radiated from distant stars, the total mass and energy of the universe stays constant.

How much matter is in the universe? We have estimated it to be 5 trillion times the mass of one trillion suns, which is twice the estimated mass within the observable Yilmaz universe. Alternatively, we could assume that our universe is infinite. The universe has always had the same amount of matter-plus-energy, and the total does not change with time.

> *Thus our picture portrays an enormous universe having a constant size. We may never be able to determine whether our universe is finite or infinite. The universe has always existed, appearing approximately like we see it today, and will always remain that way. Yet the universe constantly changes in detail because of the creation of new matter. The matter from old dead stars is radiated in the form of gravitational wave energy. The gravitational and electromagnetic energy radiated from stars is converted into diffuse matter throughout space. The creation of diffuse matter keeps the universe continually changing, so that the universe remains eternally young even though it is infinitely old.*

This is our picture of the universe. It is not a product of speculation. It evolved quantitatively from the Yilmaz theory of gravity, which has a rigorous mathematical foundation and is a direct refinement of the Einstein General theory of Relativity. I personally find this universe picture to be warmly consistent with my religious beliefs. Those who reject religion should find it to be philosophically satisfying.

Our picture envisions a universe of infinite age. The universe has always existed and will always continue to exist. Five billion years ago, nuclear fusion was ignited in a collapsing cloud of hydrogen, and our sun was born. *"Let there be Light, and there was Light"*. ***That event was the moment of Creation as far as mankind is concerned.*** Our solid earth was created 700 million years later. Over billions of years life developed on earth to produce the world that we know today.

Appendix A

Density of Matter in the Universe

Luminosity Density of the Universe

Narlikar [18] (p. 304) gives in his Eq. 9.19 the following measured luminosity density of the universe, which was derived from the *Revised Shapley-Ames Catalog*:

$$\text{Luminosity density} = 2.18 \times 10^8 \, L_{sun} h_0 / Mpc^3 \qquad (A-1)$$

where L_{sun} is the luminosity of the sun and Mpc means million parsecs, which is equal to 3.26 MLyr (3.26 million light-years). Our assumed Hubble constant H_0 is 20 km/sec per MLyr, which is 65 km/sec per Mpc. Dividing this by 100 km/sec per Mpc gives a normalized Hubble constant h_0 of 0.65. Hence Eq. A-1 can be expressed as

$$\text{Luminosity density} = 4.09 \times 10^6 \, L_{sun} / MLyr^3 \qquad (A-2)$$

Dark Matter

Measurements of the motions of galaxies and clusters of galaxies show that there must be much more dark matter (which we cannot see) than there is luminous matter (which we can see). Otherwise these groups of stars would fly apart. Narlikar [18] (p. 310) gives in his Table 9.1 the data shown in the first data column of Table A-1. These values are expressed in terms of the normalized Hubble constant h_0. The last column gives the values for the ratio η that correspond to our normalized Hubble constant h_0, which is 0.65. The parameter η in Table A-1 is the ratio of total mass to luminous mass.

In Table A-1, items (1) to (4) apply to the rotational motions of galaxies. These data suggest that the total mass associated with the rotation of a single galaxy is about 8 times the luminous mass.

Items (5) to (8) involve motions of groups of galaxies. The mass ratio is much greater for a galaxy cluster than for a single galaxy. The rotation of a single galaxy involves matter in the vicinity of the galaxy,

A. Density of Matter in the Universe

whereas the rotation of a cluster involves the total matter of the cluster.

The average distance between galaxies is about 10 million light-years (10 MLyr), and so we can allocate to each galaxy a volume of (10 MLyr)^3, which is 1000 MLyr^3. This volume is about 5 million times greater than the volume of the galaxy itself. Even though the density of dark matter is much smaller in the intergalactic space between galaxies, than in the vicinity of a galaxy, the total dark matter in intergalactic space is much greater than the dark matter close to the galaxy.

Table A-1: Average ratio (η) of total mass per luminous mass

Object	η/h_0	η
(1) Our Galaxy (inner part)	6 ± 2	3.9 ± 1.3
(2) Our Galaxy (outer part)	40 ± 30	26 ± 20
(3) Spiral galaxies	9 ± 1	5.9 ± 0.7
(4) Elliptical galaxies	10 ± 2	6.5 ± 1.3
(5) Galaxy pairs	80 ± 20	52 ± 13
(6) Local Group	160 ± 80	104 ± 52
(7) Statistics of clustering	500 ± 200	325 ± 130
(8) Abell clusters	500 ± 200	325 ± 130

Item (5) involves a pair of galaxies, and item (6) for our local group involves a small galaxy cluster. In contrast, items (7), (8) involve large clusters of galaxies, and so should give a better indication of the effects of intergalactic mass. Items (7), (8) show that the mass of intergalactic dark matter is 325 times the luminous mass of the galaxy itself. ***Therefore intergalactic dark matter (which we cannot see) is about 325 times greater than the luminous matter (which we can see).***

The Source of Dark Matter

What is the source of dark matter? Big Bang theorists are searching hard for dark matter, because their theories have difficulty explaining the early evolution of the universe unless the density of matter is close to the *critical mass density*. (See page 158, line 7.) They have not found sufficient dark matter in the universe to achieve critical mass density.

A fundamental mistake has been made in the Big Bang search for dark matter. Silk [21] (p. 163) shows that quasar radiation has been used to measure the density of intergalactic hydrogen. Since quasars are assumed to be at enormous distances, the density of intergalactic hydrogen is believed to be extremely small. However, Arp showed that quasars are relatively close, and so the estimates of intergalactic hydrogen derived from quasar radiation are not meaningful. Besides, most of the hydrogen in space is probably molecular hydrogen (H_2), rather than atomic hydrogen (H), and molecular hydrogen is very difficult to detect.

Dark matter probably consists primarily of hydrogen atoms in the enormous spaces between galaxies. A frantic search for exotic missing dark matter is being performed by some astronomers. This is a consequence of the lack of open debate in astronomy today. If astronomers listened to Halton Arp, they would recognize that quasars are probably close, and so they would not use quasar tadiation to measure the density of intergalactic hydrogen.

Total Mass in the Universe

For our assumed Hubble constant of 20 km/sec per MLyr, the radius r_0 of the observable universe is 15,000 MLyr (15 billion light-years). The volume of the observable Big Bang universe is

$$\text{Universe volume} = (4/3)\pi r_0^3 = 14.1 \times 10^{12} \text{ MLyr}^3 \qquad (A\text{-}3)$$

Multiplying this by the luminous density of the universe in Eq. A-2 gives the total luminosity of the universe in equivalent suns, which is

$$\text{Luminosity of universe} = 5.77 \times 10^{19} \, L_{sun} \qquad (A\text{-}4)$$

Stellar mass is approximately proportional to luminosity for a large collection of stars. Hence if the sun luminosity L_{sun} is replaced by the sun mass M_{sun}, Eq. A-4 gives the approximate mass of the luminous matter. Multiplying this by 325 to account for dark matter gives the following for the total mass of the observable universe

$$\text{Mass of observable universe} = 18.8 \times 10^{21} \, M_{sun} \qquad (A\text{-}5)$$

Silk [20] (p. 396) reports that the typical distance between galaxies is 10 million light-years (10 MLyr), and so the average volume of space per galaxy is 1000 MLyr^3. Dividing the universe volume in Eq. A-3 by 1000 MLyr^3 gives the following for the number of galaxies in the universe:

A. Density of Matter in the Universe

$$\text{Galaxies in universe} = 14.1 \times 10^9 \qquad (A\text{-}6)$$

Multiplying the luminosity density of Eq. A-2 by the average volume per galaxy (1000 MLyr^3) gives the average luminosity per galaxy:

$$\text{Luminosity per galaxy} = 4.09 \times 10^9 \, L_{sun} \qquad (A\text{-}7)$$

Mass Density of the Universe

Since stellar mass is approximately proportional to luminosity for a large collection of stars, we can replace L_{sun} by M_{sun} in Eq. A-2 to give the following for the density of luminous matter

$$\text{Density of Luminosity matter} = 4.09 \times 10^6 \, M_{sun}/\text{MLyr}^3 \qquad (A\text{-}8)$$

Multiplying this by the dark matter ratio 325 gives the total density of matter:

$$\text{Total density of matter} = 1.33 \times 10^9 \, M_{sun}/\text{MLyr}^3 \qquad (A\text{-}9)$$

Replace M_{sun} by the sun mass, 1.99×10^{33} gram. Since one light-year (Lyr) is 9.46×10^{12} km, MLyr is 9.46×10^{21} meters. This gives the following for the total mass density of the universe:

$$\text{Mass density} = 3.125 \times 10^{-24} \text{ grams/meter}^3 \qquad (A\text{-}10)$$

The mass of a hydrogen atom is 1.67×10^{-24} gram, and so this mass density is equivalent to 1.87 hydrogen atoms per cubic meter. *This shows that our best estimate of the average mass density of the universe is 1.9 hydrogen atoms per cubic meter.*

Predicted Density of Matter

The Yilmaz cosmology model predicts an average density of matter in the universe of $(3/8\pi G T_0^2)$, where T_0 is the apparent universe age and G is Newton's gravitational constant (6.674×10^{-8} cm^3/gm-sec^2). There are 31.558 million seconds per year, and so the apparent universe age T_0 (15 billion years) is 4.734×10^{17} seconds. The above formula gives an average density of matter of 7.98×10^{-30} gm/cm^3 or 7.98×10^{-24} grams per cubic meter. A hydrogen atom has a mass of 1.67×10^{-24} gm, and so this density is equivalent to 4.78 hydrogen atoms per cubic meter.

Thus the Yilmaz theory predicts an average density of matter in the universe equivalent to 4.8 hydrogen atoms per cubic meter. When one considers the great errors involved in these calculations, this predicted density is remarkably close to the 1.9 hydrogen atoms per cubic meter density derived from astronomical measurements.

Big Bang theories define a *critical mass density* for the universe. If the density of matter is less than critical, the universe should expand forever; and if the density is greater than critical the universe should eventually collapse. The critical mass density for the Big Bang theory has the same value $(3/8\pi G T_0^2)$ that is required by the Yilmaz theory, and so is also equivalent to 4.8 hydrogen atoms per cubic meter.

Theoretical Mass of the Universe

Thus, the Yilmaz cosmology model predicts a mass density of the universe of $(3/8\pi G T_0^2)$, which is also the critical mass density for the Big Bang universe. Multiply this mass density by the volume of the observable universe, $(4/3)\pi r_0^3$, where r_0 is the radius of the observable universe. This gives a total mass of $(r_0^3/2GT_0^2)$. The ratio (r_0/T_0) is equal to the speed of light c, and so the total universe mass becomes $(c^2 r_0/2G)$. The normalized relativistic mass of the sun, denoted (m_{sun}), is defined as $(M_{sun}G/c^2)$. Hence the total mass of the observable Big Bang universe for critical mass density is equal to

$$\text{Mass of observable universe} = (c^2 r_0/2G) = M_{sun}(r_0/2m_{sun}) \quad (A\text{-}11)$$

The normalized mass of the sun m_{sun} is 1.475 km. Since one light-year Lyr is 9.46×10^{12} km, the radius of the observable universe r_0, which is 15 billion light-years, is 142×10^{21} km. Substituting these values into the above equation gives 48×10^{21} M_{sun} for the universe mass.

For critical mass density, the observable Big Bang universe has 48×10^{21} times the sun mass. This is also the theoretical mass of the Yilmaz cosmology model for a sphere with a radius of 15 billion light-years. Equation A-5 gives a measured mass for the observable universe of 18.8×10^{21} times the sun mass, which is about 40 percent of the theoretical mass. When we consider the large errors involved in these calculations, this agreement between theoretical and measured values is very good.

A. Density of Matter in the Universe   159

Rate of Creation of Matter

Narlikar [18] (p. 240, Eq. 8-4) shows that the rate of mass creation to compensate for the Hubble expansion is $(3\rho/T_0)$. Setting ρ equal to 4.8 hydrogen atoms per cubic meter shows that 0.96 (approximately 1.0) hydrogen atom is created per cubic meter every billion years, or one hydrogen atom is created every year within a cubic kilometer.

Our earth, with a radius of 6378 km, has a volume of 1.09×10^{12} cubic kilometers, and so 1.09×10^{12} hydrogen atoms would be created per year within a volume the size of the earth, which comes to 34,500 hydrogen atoms per second. The equivalent energy (Mc^2) of the hydrogen atom (1.67×10^{-27} kilogram) is 1.50×10^{-10} watt-second. Hence 34,500 hydrogen atoms per second is equivalent to 5.18×10^{-6} watt (5.18 microwatt). *This shows that the predicted rate of creation of matter is equivalent to the continual conversion of 5 microwatts of power into matter within a volume the size of the earth.*

Size of the Initial Big Bang Universe

A star with our sun's mass but the density of water would have a radius of 780,000 km, obtained by multiplying 700,000 km (sun radius) by the cube root of 1.4 (sun density). Multiply 780,000 km by the cube root of 18.8×10^{21} (from Eq. A-5) to obtain the radius for the observable universe if it were squeezed into a single body with the density of water:

Universe radius = 20.7×10^{12} km = 2.19 Lyr   (water density)   (A-12)

One light-year (Lyr) is 9.46×10^{12} km. The density of a neutron star is 2.0×10^{14} times the density of water. The cube root of this ratio is 58,480. Hence the radius of Eq. A-12 is divided by 58,480 to obtain the following initial radius of the observable universe for neutron-star density:

Universe radius = 354 million km   (neutron-star density)   (A-13)

This is 1.55 times the radius (228 km) of the orbit of Mars around the sun. *From astronomical data, the observable Big Bang universe would have had an initial radius 1.55 times the radius of the Mars orbit if it began with the density of a neutron star, the maximum possible density of matter.*

Appendix B

Cosmic Microwave Background Radiation

Big Bang proponents claim that cosmic microwave background radiation validates their theory. However, the Yilmaz cosmology model also predicts this radiation. In Chapter 12, Fig. 12-4 shows that at very large distances the apparent galaxy velocity is very close to the apparent speed of light, and so the light radiated from distant galaxies should be Doppler shifted to very low frequencies. *Believe* [1] presents an analysis of this radiation in Appendix D, which is summarized in this appendix. This analysis is also presented in the *Website* [3], Page 4. In this appendix the plots have been modified to correspond to a Hubble constant of 20 km/sec per million light-years (rather than 25).

This cosmic microwave radiation predicted by the Yilmaz cosmology model is equivalent to the emission from an ideal blackbody at a temperature between 2.1 and 3.4 °K. This agrees closely with the 2.73 °K blackbody temperature measured by the COBE satellite.

The general spectrum of blackbody radiation was shown in Fig. 3-1 of Chapter 3. This spectrum is expressed in terms of the half-power frequency f_h of the spectrum. The equivalent half-power wavelength λ_h, is equal to c/f_h. This wavelength is related as follows to the blackbody temperature T expressed in degrees Kelvin:

$$\lambda_h = 4.107/T \text{ millimeter (mm)}$$

The temperature of a blackbody determines the actual frequencies that it radiates, but the shape of the spectrum is the same for all temperatures.

The light radiated from our sun has a spectrum approximating that of an ideal blackbody at a temperature T of 5770 °K. The corresponding value for the wavelength λ_h is 0.712 micrometers (millionths of a meter). Our analysis approximates the spectra of all stars in our universe with that of our sun.

Because the light from a galaxy is Doppler shifted toward lower frequency, the equivalent blackbody temperature of the spectrum that is

B. Cosmic Microwave Background Radiation   161

received from a galaxy decreases with distance to the galaxy. Figure B-1 shows the equivalent blackbody temperature of the Doppler-shifted spectrum that is received from a galaxy at a particular true distance. This was calculated by applying the Einstein formula for Doppler frequency shift (given in *Story* [4], Appendix E) to the data in Fig. 12-4 of Chapter 12, which shows the apparent galaxy velocity divided by the apparent speed of light for the Yilmaz cosmology model. It is the apparent velocity ratio that determines the Doppler frequency shift.

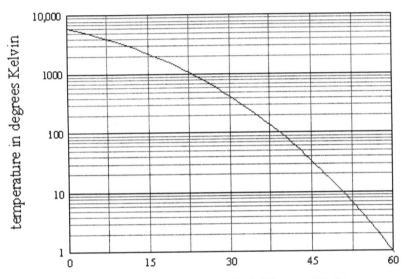

Figure B-1: Blackbody temperature of radiation received from a distant galaxy versus the true distance r to the galaxy

In Fig. B-1 the equivalent blackbody temperature decreases from 5770 °K (the blackbody temperature of the sun) for a close galaxy down to 1 °K for a galaxy at a true distance of 60 billion light-years.

Figure 12-3 of Chapter 12 showed that the apparent density of matter should be very high as the apparent distance to a galaxy approaches the limit of 18.8 billion light-years. Because of this very high apparent density, the intensity of radiation received from a galaxy located at a large true distance should also be very large. Figure B-2 shows the photon rate that should be received per unit area of receiver surface, from galaxies at different values of true distance. The plot shows that the photon rate becomes extremely large at large true distances.

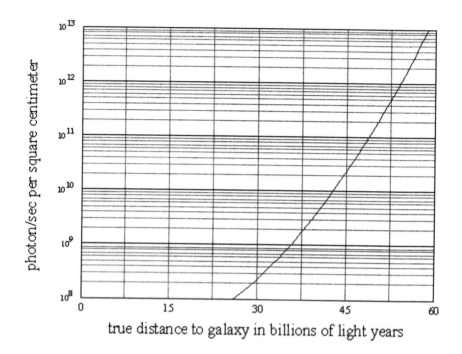

Figure B-2: Photon rate per unit area for the cosmic radiation received from galaxies at a true distance r.

By combining the data in Figs. B-1 and B-2, we can plot the intensity of the received radiation versus the equivalent blackbody temperature of the spectrum. This gives the solid curve in Fig. B-3. This shows the photon rate that would fall onto a square centimeter of receiver surface, expressed in terms of the equivalent blackbody temperature T of the received Doppler-shifted spectrum.

An ideal blackbody emits a photon rate that is proportional to the cube of the blackbody temperature. The dashed plot in Fig C-3 shows the photon rate per unit area that is emitted from an ideal blackbody.

For an ideal blackbody, the radiation is in thermal equilibrium with the molecules at the surface. We assume that this radiation level cannot be exceeded by cosmic radiation in space. If it were, the diffuse material in space should rapidly absorb the cosmic radiation. Therefore we conclude that the Doppler-shifted galaxy radiation indicated by the solid plot in Fig C-3 cannot exceed the dashed plot. This indicates that the intersection point of the two plots should give the effective blackbody temperature of the received blackbody radiation. The figure shows that this calculated temperature is 2.7 °K. Because of approximations in the analysis, there can be an error in this result. We conservatively estimate that an exact computed temperature should fall within the range from 2.1 °K to 3.4 °K.

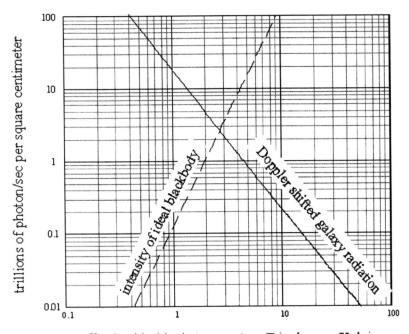

effective blackbody temperature T in degrees Kelvin

Figure B-3: Photon rate intensity of cosmic radiation received from galaxies, versus effective blackbody temperature of radiation; received radiation (solid); ideal blackbody (dashed)

Figure B-1 shows that galaxies producing blackbody radiation equivalent to a 2.7 °K temperature are at a distance of 56 billion light-years. This indicates that we should receive from galaxies at a true distance of about 56 billion light-years cosmic microwave radiation that corresponds to a blackbody at a temperature from 2.1 °K to 3.4 °K.

For comparison, the COBE satellite found that the received cosmic radiation is equivalent to the radiation from an ideal blackbody at a temperature of 2.73 °K. The COBE radiation is received with very high uniformity from all directions, which agrees with the Yilmaz cosmology model.

GLOSSARY

numbers, exponential representation
   10^5 means 1 followed by 5 zeros; 10^{-5} means $1/10^5$.
   For example, $3.1 \times 10^5 = 3.1 \times 100{,}000 = 310{,}000$.
   $3.1 \times 10^{-5} = 3.1/100{,}000 = 0.000{,}031$

number prefixes:
   nano = billionth ($1/10^9$)   mega = million (10^6)
   micro = millionth ($1/10^6$)   kilo = thousand (1000)
   milli = thousandth (1/1000)
one billion = 1000 million; one trillion = 1000 billion

acceleration: The rate-of-change of velocity, relative to time.

acceleration of gravity g: Rate of increase of velocity when a body is allowed to fall to earth; equal to 9.8 meter/second per second at earth surface.

aether (also called ether): Postulated medium that provides an absolute reference relative to which light propagates; Einstein rejected the *aether* concept.

apparent age of the universe (T_0): The time elapsed since the Big Bang if the universe has always expanded uniformly at the present Hubble constant; equal to 15 billion years for a Hubble constant of 20 km/sec per million light-years.

archaea: Cells without nuclei that tolerate extreme environment; not bacteria.

Big Bang theory: The hypothesis that our universe began as a highly dense mass that exploded with a Big Bang about 15 billion years ago.

black dwarf star: The dark ember of a white dwarf star after it stops radiating energy; the ultimate fate of our sun.

black hole: Einstein theory seems to predict that a star with a mass-to-radius ratio 240,000 times our sun collapses to form a *black-hole singularity* of zero size and infinite mass density; light cannot escape the *event horizon* around a black hole.

blueshift: Wavelength shift to shorter wavelength, defined as wavelength decrease divided by normal wavelength; proportional to star velocity toward the earth.

Cepheid variable star: A star that varies periodically in power; radiated power varies with frequency, and so this star gives a measure of its distance.

chlorophyll, the chemical in cells that implements *photosynthesis.*

coordinates: For 3-dimensional spatial measurements: *rectangular (or Cartesian) coordinates* use 3 distances;. *Relativity coordinates* have a fourth dimension that represents time.

cosmology: The study of the evolution and large-scale structure of our universe.

covariance, Principle of: Expressing a law of physics so it is independent of the velocity, acceleration and gravitational field at the location of the observer.

diagonal elements of a tensor: Those 4 elements for which the two indices are equal; the other 12 elements are called *nondiagonal.*

Doppler wavelength shift: Wavelength change due to velocity; approximately equal to normal wavelength times radial velocity divided by speed of light.

earth parameters: radius = 6378 km; mass = 5.9742×10^{24} kg; orbit radius = 149.6 million km.

Einstein gravitational field equation: Tensor formula specifying the Einstein General theory of Relativity; which represents 10 independent equations, which relate the energy-momentum tensor to the Ricci tensor.

electromagnetic field equations: Mathematical formulas derived by Maxwell, which specify the physical laws of electricity and magnetism.

electromagnetic wave: Light wave, radio wave, etc., which consists of electric and magnetic fields oscillating at right angles to one another.

electron: See elementary particles of atom.

elementary particles of atom: Atom consists of: (1) *electron* with negative electrical charge, (2) positive charged *proton*, with 1836 times electron mass, and (3) *neutron* with no charge and approximately mass of proton.

equivalent energy of matter: Energy equal to Mc^2 that would be released if a mass M were converted into energy; 25 million kilowatt hours of energy per gram.

ether: See aether.

eukaryote: Cell with nucleus; all multi-celled organisms consist of eukaryote cells.

event horizon: Spherical surface surrounding a *black-hole* over which speed of light is zero; light theoretically cannot escape from within this surface.

galaxy: A group of many billons of stars; nearly all of the stars of the universe are parts of galaxies; but some are parts of smaller groups called *stellar clusters*.

galaxy, types of: Spiral galaxy is like Milky Way galaxy, with disk shape; other major type is the *elliptical galaxy*, with only an elliptical nucleus.

General theory of Relativity: Generalization (1916) of Einstein's Relativity theory to include effects of acceleration and gravity.

gravitational constant G: Constant of proportionality in Newton's law of gravitational attraction; 6.674×10^{-8} cm^3/gm-sec^2.

Hubble Constant: The ratio of receding galaxy velocity to galaxy distance for distant galaxies; the latest value is about 20 km/sec per million light-years.

Hubble expansion: The 1929 finding by Hubble that galaxies are receding at velocities approximately proportional to distance.

Hubble Law: The postulate that our whole universe expands at a constant rate.

Light-year: Distance light travels in one year; 9.46 trillion kilometers.

Lorentz transformation: Special Relativity formulas derived by Lorentz, giving distance and time differences measured by observers at different velocities.

magnitude: Logarithmic measure of power received from a star; power decreases by 2.5 for a magnitude change of +1, and by 100 for a magnitude change of +5.

magnitude, absolute: Magnitude a star would have if observed at a distance of 32.6 light-years (10 parsecs); our sun has absolute magnitude of 4.8.

Marmet redshift: The prediction by Paul Marmet that a hydrogen cloud produces a redshift proportional to gas density and cloud thickness.

mass, rest: Mass of object varies with velocity; *rest mass* is mass at zero velocity.

metric ton, 1000 kilograms or one million grams, equal to 2205 pounds, or about 10% more than an English ton (2000 pounds).

momentum: The product of mass times velocity.

momentum, angular: For mass elements rotating about a center of gravity, the sum (for each element) of momentum times radial distance from the center.

moon parameters: radius = 1738 km, mass = 7.35×10^{22} km, distance = 384,000 km

nebula: Fixed astronomical object with extended image; now restricted to gas cloud heated by star radiation, called *gaseous nebula.*

neutron: See elementary particles of atom.

neutron star: A star consisting of tightly packed neutrons; having density of matter in atomic nucleus, 200 million tons per cubic centimeter.

nucleus of atom: Extremely compact body holding protons and neutrons; having density of 200 million tons per cubic centimeter.

nuclear fission: Energy released by splitting heavy atomic nucleus; operating in atomic nuclear bomb and nuclear power plant.

nuclear fusion: Energy released by fusing light atomic nuclei; operating in the sun and the hydrogen bomb.

parallax: Relative image shift of a nearby object due to motion of the observer.

parsec: Theoretical distance of a star exhibiting an annual parallax shift of ±1 arc second because of rotation of the earth around the sun; equal to 3.26 light-years.

photosynthesis: The process in plants implemented by *chlorophyll,* which uses the energy from sunlight, the hydrogen in water, and the carbon in carbon dioxide to synthesize carbohydrate food, and releases oxygen as a byproduct.

proper coordinates: Proper coordinates move with an object; *proper distance* and *proper time* are measurements made on the object relative to these coordinates.

proton: See elementary particles of atom.

pulsar: A star that emits radio pulses at precise intervals, typically at 10 to 1000 pulses per second; a rapidly spinning neutron star.

quasar or quasi-stellar object (QSO): A star-like object with a very large redshift.

radial velocity, component of velocity of a body, along a radial line from the observer to the body; Doppler wavelength of a star is approximately proportional to the radial velocity of the star relative to the earth.

red giant star: State our sun will reach in 5 billion years after converting nearly all of its hydrogen to helium; sun will swell to half the distance to the earth.

receding velocity: Velocity component away from earth.

redshift: Wavelength increase of spectral lines divided by normal wavelength.

redshift, Doppler: Redshift due to velocity; approximately equal to receding velocity divided by speed of light.

redshift, gravitational: Redshift produced by a gravitational field.

redshift, intrinsic: Component of galaxy or quasar redshift unrelated to velocity.

Relativity, Special theory of: Einstein's basic theory (1905), which shows how time and spatial measurements are changed by the velocity of an observer.

Schwartzschild limit: The maximum mass-to-radius ratio of a star for which the Schwartzschild solution has a real answer; 240,000 times the ratio for our sun.

Schwartzschild solution: First analytical solution of Einstein General Relativity theory (1916), applied to star with constant density and no viscosity.

singularity: Big Bang prediction in which size shrinks nearly to zero without change of mass, so that density of matter is nearly infinite.

solar system: the planets, comets, and other bodies orbiting a star.

spectral lines: In addition to a continuous spectrum, star light has bright emission lines and dark absorption lines at discrete wavelengths.

spectrum: Pattern formed by passing light through a prism to separate wavelengths.

speed of light c: 300,000 kilometers per second (exactly 299,793 km/sec).

stars, types of (See individual types)
  black dwarf, neutron star, pulsar, red giant, white dwarf.

Steady-State Universe theory: In 1948 Fred Hoyle, et al, postulated that universe age is infinite, and diffuse matter is created to compensate for Hubble expansion.

stellar: pertaining to a star.

supernova, An enormous stellar explosion equivalent to billions of suns.

sun parameters: radius = 696,000 km; mass = 1.989×10^{30} kg; density = 1.4; normalized mass m = 1.475 km

Tangential velocity of a body, the component of velocity of a body that is perpendicular to a radial line drawn from the observer to the body.

temperature, scales of: Celsius (or Centigrade) scale is zero °C at freezing point of water and 100 °C at boiling point; Kelvin scale has Celsius intervals but is zero at absolute zero (-273.15 °C), where random molecular motion is zero.

tensor definition: A generalized variable, usually with 16 elements in Relativity theory; but the high-order Riemann tensor has 256 elements.

tensor, types of :

  Einstein tensor, G_{ab}, describes the curvature of space; closely related to the Ricci tensor.

  energy-momentum tensor: T_{ab}, describes the properties of matter and energy.

  metric tensor: g_{ab}, describes the shortest distance between two points in curved space.

  Ricci tensor: R_{ab}, describes the curvature of space; closely related to Einstein tensor, and derived from Riemann tensor.

  Riemann tensor, Uniquely describes curvature of space; has 4 indices and 256 elements; not used directly in Relativity calculations.

  stress-energy tensor for gravitational field: The Yilmaz tensor that specifies energy and stress of gravitational field (not used in Einstein theory).

tensor, diagonal: This class of tensor has only 4 nonzero elements, which are on the diagonal of the tensor matrix (both indices are equal).

ton, See *metric ton*

vector: Variable having amplitude and direction, represented by an arrow.

velocity, radial or tangential: Radial velocity is velocity component along a radial line from earth to a star; toward or away from earth; *tangential velocity* is perpendicular to a radial line.

white dwarf: When nuclear fuel is depleted, our sun will shrink to a *white dwarf*, after reaching size of the earth, it will cool to become a *black dwarf*.

Yilmaz theory of gravity: Refinement of the Einstein General theory of Relativity, which has achieved an exact solution to the principles of the Einstein theory.

BIBLIOGRAPHY

There are two sets of references. The preface Y indicates references on the Yilmaz theory.

Yilmaz Theory Bibliography

[Y1] Huseyin Yilmaz, "New Approach to General Relativity", *Physical Review*, vol. 111, No. 5, Sept. 1, 1958, pp 1417-1426,

[Y2] Huseyin Yilmaz, "New Theory of Gravitation", *Physical Review Letters*, vol. 27, No. 20, 15 Nov. 1971, pps. 1399-1402.

[Y3] Huseyin Yilmaz, "New Approach to Relativity and Gravitation", *Annals of Physics*, Academic Press, NY, 1973, pps. 179-200.

[Y4] Huseyin Yilmaz, "New Theory of Gravitation", *Proc. 4th Marcel Grossman Meeting Gen. Relativity*, Remo Ruffini, ed, Rome Univ, Italy, June 1985.

[Y5] Huseyin Yilmaz, "New Direction in Gravity Theory", *Hadronic Journal*, 1986, vol. 9 No 6, pp 281-291,.

[Y6] Huseyin Yilmaz, "Present Status of Gravity Theories", *Hadronic Journal*, 1986, vol. 9 No 6, pp 233-238.

[Y7] Huseyin Yilmaz, "Dynamics of Curved Space", *Hadronic Journal*, 1986, vol. 9 No 2, pp 55-60.

[Y8] H. Yilmaz, "Toward a Field Theory of Gravitation", *Nuovo Cimento, B Gen. Physics*, 1992, vol. 107, Iss. 8, pp 941-960.

[Y9] Yilmaz, Huseyin, "Did the Apple Fall?", in *Frontiers of Fundamental Physics*, 1994, M. Barone and F. Selleri, eds, pp. 115-124, Plenum Press, NY.

[Y10] Alley, Carroll O., "Investigation with lasers, atomic clocks [etc.] of gravitational theories of Yilmaz and Einstein", in *Frontiers of Fundamental Physics*, 1994, M. Barone and F. Selleri, eds, pp. 125-137, Plenum Press, NY.

[Y11] Huseyin Yilmaz, "Gravity and Quantum Field Theory, a Modern Synthesis", *Ann New York Acad Science*, 1995, vol 755, pp 476-499.

[Y12] Carroll O. Alley, "The Yilmaz Theory of Gravity and its Compatibility with Quantum Theory", *Ann New York Acad Science*, 1995 vol 755, pp 464-477.

General Bibliography

[1] Adrian Bjornson, *A Universe that We Can Believe*, Addison Press, Woburn, MA, 2000, ISBN 09703231-0-7.
[2] Adrian Bjornson, *How Was Our Universe Created?*, Addison Press, Woburn, MA, 2000, ISBN 09703231-1-5.
[3] Internet Website www.olduniverse.com.
[4] Adrian Bjornson, *The Scientific Story of Creation*, Addison Press, Woburn, MA, 2000, ISBN 09703231-2-3.

Textbooks on Relativity

[5] Albert Einstein, *The Meaning of Relativity*, Princeton University Press, 5th ed., 1953, ISBN 0-691-02352-2, (1st ed. 1921), (*See* appendix for 2nd ed., 1945, p. 129).
[6] Tullio Levi-Civita, *The Absolute Differential Calculus*, 1977, Dover Pub, NY, (Italian ed, 1923), ISBN 0-486-63401-9.

Classic Papers on Relativity

[7] H. A. Lorentz, "Electromagnetic Phenomena in a System Moving with any Velocity Less than that of Light", *Proc. Acad Science Amsterdam*, 1904, vol 6, reprint in *The Principle of Relativity*, Dover Pub, NY, 1952, pp. 9-34.
[8] A. Einstein, "On the Electrodynamics of Moving Bodies", Annalen der Physik, 1905, v. 17, English trans. in *The Principle of Relativity*, 1952, Dover Pub, NY, pp. 35-71.
[9] Albert Einstein, "The Foundation of the General Theory of Relativity", Annalen der Physik, vol. 49, 1916, English transl. in *The Principle of Relativity*, 1952, Dover Pub, NY, pp. 109-164.
[10] G. Ricci and T. Levi-Civita, "Methods de calcul differential absolu et leurs applications", *Math. Ann.*, 1901, vol. 54, pp. 125-201.
[11] J. R. Oppenheimer and H. Snyder, "On Continued Gravitational Contraction", *Physical Review*, Sept. 1939, vol 56, pp 455-459.
[12] Albert Einstein, "On a stationary system with spherical symmetry consisting of many gravitating masses", *Annals of Mathematics*, Oct. 1939, vol 40, No 4, pp 922-936.

Books on Cosmology

[13] George Gamow, *One, Two, Three . . Infinity*, Bantam Books, 1967, 1st ed. 1947.
[14] Halton C. Arp, *Quasars, Redshifts, and Controversies*, 1987, Interstellar Media, Berkeley, Calif, ISBN 0-941325-00-8.
[15] Halton C. Arp, *Seeing Red*, 11998, Aperion, Montreal, Quebec, ISBN 0-9683689-0-5. (available at Internet website *www.Amazon.com*)
[16] Eric Lerner, *The Big Bang Never Happened*, Times Books div Random House, NY, 1991, ISBN 0-8129-1853-3.
[17] David Filkin, *Stephen Hawking's Universe, the Cosmos Explained*, 1997, Basic Books div Harper Collins, NY, ISBN 0-465-08199-1.
[18] J.V. Narlikar, *Introduction to Cosmology*, 1993, 2nd Ed., Cambridge U. Press, Cambridge, England, ISBN 0-521-42352-X.
[19] Fred Hoyle, Geoffrey Burbidge, and Jayant Narlikar, *A Different Approach to Cosmology*, 2000, Cambridge U. Press, England, ISBN 0-521-*66223-0*.
[20] Joseph Silk, *The Big Bang*, 1989, W. H. Freeman, NY, ISBN 0-7167-1812-X.
[21] Joseph Silk, *A Short History of the Universe*, 1994, Scientific American Library, W. H. Freeman, NY, ISBN-0-7167-5048-1.
[22] John A. Peacock, *Cosmological Physics*, 1999, Cambridge U. Press, United Kingdom, ISBN 0-521-42270-1.

Book on Albert Einstein
[23] Albrecht Folsing, *Albert Einstein, a Biography,* 1997, (trans. from German by Ewald Osers). Penguin Books, NY, ISBN 0-14-02.3719-4.

Books on Astronomy
[24] Terrence Dickinson, *The Universe and Beyond,* 3rd ed, 1999, Firefly Books Ltd., Ontario, ISBN 1-55209-361-1.

Papers on Astronomy and Cosmology
[25] Geoffrey Burbidge, "Why Only One Big Bang", *Scientific American,* Feb. 1992 , p. 120.
[26] Peebles, Schramm, Turner, and Kron, "The Evolution of the Universe", *Scientific American,* Oct. 1994, pps 53-65.
[27] Ann Finkbeiner, Astronomy: Hubble Telescope Settles Cosmic Distance Debate, or Does it?", *Science,* May 28, 1999.
[28] Paul Marmet, "A New Non-Doppler Redshift", presented in Internet website: www.newtonphysics.on.ca.
[29] S. Chandrasekhar, "The Dynamic Instability of Gaseous Masses Approaching the Schwartzschild Limit in General Relativity", *Astrophysical Journal,* vol. 140, No. 2, 1964, pp. 417-433.
[30] Ivars Peterson, "A New Gravity: Challenging Einstein's general theory of relativity", *Science News* , Dec. 3, 1994, vol. 146, pps 376-378.

Gravity and Light (Chapters 4 and 5)
[31] Will and Ariel Durant, *The Age of Reason Begins,* 1961, MJF Books, NY, ISBN 1-56731-018-4, pps 584-611.
[32] Will and Ariel Durant, *The Age of Louis XIV,* 1963, MJF Books, NY, ISBN 1-56731-019-2, pps 531-547.
[33] Hal Hellman, *Great Feuds in Science,* 1998, John Wiley, NY, ISBN 0-471-16980-3, Ch. 1.

INDEX

Numbers in brackets [ ] are Bibliography references.

acceleration, 54, 56. 58-62, 64, 84-90, 102
acceleration of gravity, 54, 58, 61, 62, 64, 84, 85, 102
aether, 69-73
age of earth, 14
age of sun, 2, 14
age of stars, 118
age of universe
   apparent age, 33-34, 157
   true age, 33-35, 118-119, 153, 145-146
   age dilemma, 118-119
Alfven, Hannes, 117-122
algae, 16
Alpher, Ralph, 41
Alley, Carroll O, 123-128, [Y10, Y12]
amphibian, 17
archaea, 15-17
Arp, Halton, 46-50, 117, 156, [14, 15]
atom, structure of, 23-25

bacteria, 15-17
big bang theory, 4-7, 9-10, 12, 25, 33-38, 41-43, 104-106, 116-122, 128-130, 133-134, 136, 138, 143, 145, 148-152, 155-156, 158-160
bird, 17-18

blackbody, 39-42, 119, 143, 149, 160-163
   see also, cosmic microwave radiation
black hole, 9, 36, 42-45, 105, 110-114, 129
   see, singularity
Bondi, Hermann, 35-36
Brahe, Tycho, 52
Burbidge, Geoffrey R., 47, 116, [19, 25]

calculus, 56, 57, 61, 88, 89 ,91
calculus, absolute differential, 88, 91
Cavendish, Henry, 59, 63-64, 125
Cepheid variable stars, 29-32
Chandrasekhar, S., 48, 49, [29]
clock rate, change of,
   special relativity, 77-79
   general relativity, 85-88, 113-115,
   Yilmaz theory, 113-115
   Yilmaz cosmology model, 134-138
conservation of energy and matter, 126
coordinates, 80. 82, 89-90, 92, 95-96, 98, 124, 138
   four dimensional, 90, 96, 98
   proper, 80, 82, 138
   rectangular (Cartesian), 92
Copernicus, 51-53
cosmology theories,
   see, universe, models of
cosmic background explorer satellite, COBE, 42, 143, 149, 160, 163

cosmic microwave background
   radiation, 12, 36, 38-39, 41-42, 119,
   133, 143, 149, 160, 162-163
   see also, blackbody radiation
covariance, 89
creation of
   Biblical story, 151-153
   earth, 2, 14, 151-152
   life, 3, 14
   matter, 5, 13, 35, 142, 144, 146, 152, 159
   solar system, 2
   stars, 2, 5, 14
   sun, 2, 14, 151-152
   universe, 5, 8, 13, 129, 148, 151-153
curvature of space, 89, 96, 98

dark matter, 35, 49-50, 154-157
density of universe, 12, 157-158
density of matter, 3, 5-8, 10, 12, 21-25,
   34-37, 44-45, 49-50, 64, 99-100,
   107, 110-112, 119, 129-130, 132,
   136, 141-142, 148, 154-159, 161
Descarte, Rene, 92
diagonal tensor, 97-101, 106, 125
Dicke, Robert, 41
Dickinson, Terence, 7, 38, 41, [24]
dinosaur, 17, 18
Doppler wavelength shift, 30-31, 46,
   70, 86-87, 139, 143, 149, 160-162
dwarf star,
   white, 2, 21, 23-25, 32
   black, 2, 11, 21, 145, 147

earth, creation, 2, 14, 151-152
Einstein, Albert, 5-6, 8-11, 13, 25,
   43-45, 62, 72-75, 77, 82-85, 88-91,
   96-97, 101-106, 111-112, 122-123,
   125, 128-132, [5, 8, 9, 12, 23]
Einstein theories, 5-6, 8-11, 13, 25-27,
   37-38, 42-45, 47, 49-50, 62, 72-73,
   75-77, 82-85, 88-91, 96-97, 99-107,
   110-113, 121-133, 138, 143-145,
   152, 161

Einstein gravitational field equation,
   see, gravitational field equation
Einstein general relativity:
   single-body solution, 123-126
Einstein photo-electric effect, 73-74
Einstein relativity theories, 5, 8-11, 13,
   16, 42-44, 50, 62, 65, 68, 72-74,
   75-83, 84-101, 102-107, 111,
   122-124, 126, 128, 130-133, 135,
   138, 146, 148, 152
   Special Relativity, 13, 65, 68, 72-74,
      75-93, 102, 107, 122, 135, 138,
      146, 148
   General Relativity, 5, 8-11, 16,
      42-44, 50, 62, 84-101, 102-107,
      111, 122-124, 126, 128, 130-132,
      148, 152
Einstein *unified field theory* search,
   see, unified field theory
electron, 3, 21, 23-25, 66-67, 81-82,
   120, 144
electromagnetic, 11, 66-68, 72, 96, 98,
   106, 130, 145, 152
electric field, 66-67
energy-plus-matter conservation, 126,
   132, 148
energy-to-matter conversion, 82-83
equivalence, acceleration and gravity,
   84, 85
ether, see, aether
eukaryote, 16-17
event horizon, 45, 111

field equation, gravitational,
   see, gravitational field equation
Filkin, David, 36-37, [17]
fish, 16-17
Fischer, J. R., 118-119
FitzGerald, George, 71
Folsing, 8, [23]
Freedman, Wendy, 32
Fresnel, Augustin, 69
fusion, nuclear, 2, 9, 14, 20-21, 23, 27,
   147, 151, 153

galaxies,
 M31 and M33, 30-31
 M51 Whirlpool, 1, 26-27
 Milky Way, 1, 2, 4, 26-29
Galileo, 51, 53-54, 56-57, 120-121
Gamow, George, 5-8, 25, 36, 38, 41, [13]
Geller, Margaret, 118
geodesic, 88-90, 138, 141-142
Gold, Thomas, 35-36
gravity wave, *see,* wave, gravitational
gravitational constant G, 58-59, 63-64, 108, 125, 157
gravitational field equation, Einstein, 9, 45, 90, 96-97, 99-101, 103-106, 111-112, 121, 123, 125-128, 132
gravitational field equation, Yilmaz, 103-106
gravitational redshift,
 see, redshift, gravitational
gravitational theories,
 Einstein, see Einstein gen. relativity
 Newton, *see,* Newton
 Yilmaz, *see,* Yilmaz gravity theory

Hawking, Stephen, 36-37, [17]
helium, 2, 14, 20-21, 83
Herman, Robert, 41
Hertz, Heinrich, 68
Hoyle, Fred, 5, 11, 35-36, 47, [19]
Hubble, Edwin, 4, 30-31
Hubble expansion, 4-6, 10-13, 29-35, 119, 132-133, 136, 138-142, 149-149, 154, 156, 159-160
Hubble constant, 4, 31-35, 133, 136, 154, 156, 160
Hubble Law, 138-139
Hubble Space Telescope, 29, 32
Huchra, John P., 118
human evolution, 18-20
hydrogen, 2-3, 12, 14-15. 20-22, 30-31, 34-35, 49-50, 63, 83, 142, 144-145, 148, 151, 153, 156-159

index (indices) in tensors, 93, 95

infinite mass density, *see* singularity

Kepler, Johannes, 51-53. 56-57, 61, 121

Lerner, Eric J., 117-122   [16]
Leavitt, Henrietta, 29
Leibniz, G. Wilhelm, 57
Levi-Civita, Tullio, 88-89, [6, 10]
life on earth, 1-3, 14-17, 20, 151, 153
life of stars, galaxies, 14, 20-21, 23, 145
lifetime, Einstein, 9, 43, 45, 91, 97, 101, 106, 123, 125
light rays , bending, 90-91
light, speed of, 4, 12-13, 30, 33, 39, 44-46, 65, 67-73, 75-79, 81-82, 84, 86, 88-89, 102, 104, 106-111, 115, 119, 127, 139-140, 143, 146, 150, 158, 160-161
light, speed of,
 measurement, 68-71
 constancy of, 68-73
 variation of, due to gravity, 88, 107-113
Lorentz, Hendrick, 71-73, [7]
Lorentz transformation, 72

Magellanic clouds, 29
magnetic field, 66-67
mammals, 17-18
Marconi, Guglielmo, 68
Marmet, Paul, 34-35, 48-49, 148, [28]
Marmet redshift, *see,*
 Marmet, Paul
Mars, 6, 7, 36, 159
mass, normalized, 108-109
matrix, 95
matter and energy conservation, 126, 132, 148
matter-to-energy conversion, 82-83
Maxwell, J. Clerk, 68
Maxwell's electromagnetic
 field equations, 68, 72

174 The Mystery of Creation

measurement for star and galaxy
   distance, 27-33
   velocity, 30
Messier, Charles, 27
Mercury, 90-91, 124-125, 130
meteorite, 14-15, 151
metric tensor, *see,* tensor, metric
Michelson-Morley experiment, 70-71
microwave, 12, 16, 38-39, 41-42, 119,
   133, 143, 149, 160, 163
Milky Way, *see,* galaxy, Milky Way
momentum, 82
multi-body solution
   of Einstein theory, 123-126
myth and cosmology, 119-122

Narlikar, Jayant V., 47, 154, 159,
   [18, 19]
nebula, 1, 22, 26-27, 30, 35, 49, 120
neutrino, 22
neutron, 3, 5-7, 11, 21-22, 25, 36, 45,
   110, 112, 129, 145, 159
neutron star, 3, 5-6, 11, 21-23, 25, 36,
   45
Newton, Isaac, 50, 51-64, 80, 89, 91,
   124-125, 142, 157

observable universe
   big bang theory, 4, 6-7, 33, 36-38,
      129, 134, 136, 149-150, 156,
      158-159
   Yilmaz cosmology model, 149-150
Oppenheimer, J. Robert, 43-45,
   110-111, 123, [11]

parallax, 28-29
parsec, 28, 33, 154
Pauli exclusion principle, 25
Peacock, John A., 104-105, [22]
Peebles, P. J. E., 41, 71, 120, [26]
Penrose, Roger, 36, [17]
Penzias, Arno, 38, 41
photoelectric effect, 73-74
photosynthesis, 15-16, 151
photon, 34, 73, 138, 148, 161-162

plants, terrestrial, 17
plasma, 117, 120, 143
proton, 3, 7, 21, 23, 25, 38, 144
Proxima Centauri, 28
pseudo-tensor, 103
pterodactyl, 18
pulsar, 22

quantum mechanics, 25, 106, 130-131
quasar, 35, 45-50, 115, 155-156

radiation, blackbody, *see,* blackbody
radio, 11, 22, 46, 66, 68, 82
redshift, 4-6, 12, 30-31, 34-35, 46-49,
   85-87, 91, 102, 107, 114-115, 118,
   148
   velocity, *see,* Doppler
   gravitational, 114-115
   Marmet effect, 34-35
relativistic effects, gravitational
   clock rate, 113-114, 84-88
   Hubble expansion, 138-142
   spatial contraction, 113-114, 134-138
   speed of light, 134-135, 109-113
   wavelength, 114-115
relativistic effects, velocity, 76-79
   (clock rate, spatial contraction,
   simultaneity)
relativity theory, *see,*
   Einstein theories
   Yilmaz gravity theory
reptile, 17-18
Ricci, Gregorio, 88-89, [6, 10]
Riemann, Bernhard, 88-89

Sandage, Alan, 32
Schwartzschild, 43-45, 90, 99-100,
   104, 107, 109-114, 123-124, 126,
   128, 138
Schwartzschild, Karl, 43-44, 90
Schwartzschild limit, 44-45, 110-114,
   123, 126, 128
Schwartzschild radius, 110-111
Schwartzschild singularity, 43, 45, 111
Silk, Joseph, 37, 50, 121, 156, [20, 21]

Index **175**

singularity, 5-10, 25, 36-38. 43-45, 103, 105, 110-112, 122-123, 128-129, 130, 132
   black hole, 9, 43-45, 110-111
   Big Bang, 5-9, 36-38. 43, 122
   Einstein rejection of, 8-9, 111
   Yilmaz rejection of, 10, 105, 111
simultaneous events, 78, 80
Snyder, H., 43-45, 110-111, 123, [11]
solar system, *see*, creation of
sound, 65-66, 68-70
spatial contraction
   from gravity, 113-114
   from velocity, 76-78
   Yilmaz cosmology model, 134-138
speed of light, *see,* light, speed of
spectral lines, forbidden, 49
spectral, 4, 29-30, 34, 39, 41, 49
spectrum, 4, 5, 30, 39-42, 45, 49, 70, 87-88, 115, 119, 141, 143, 160-162
stars
   life cycles of, 20-25
   age of, 118
   distance measurement, 27-33
   velocity measurement, 30
stars, variable
   Cepheid, 29-32
   RR Lyrae, 31
   quasar variation, 46, 49
steady-state universe theory
   see, universe. models of
sun
   creation of, 2, 14, 151-152
   life and death of, 14-15, 20-21
   normalized mass, 108
supernova, 3, 14, 21-22, 32, 47

telescope, 4, 26, 28-30, 32, 45, 49, 53, 117
temperature, 14-15, 20-21, 39-42, 49, 143, 147, 149, 160-163
tensor analysis, 88-89
tensor, basic concept, 91-96
tensor, diagonal, 97-101, 106, 125

tensor, types of
   Einstein, *(variation of Ricci tensor)*
   energy-momentum, 96-101, 103, 126
   metric, 96-102, 106-107, 125
   pseudo-tensor, 103
   Ricci, 96-100, 103
   energy of gravitational field, 103, 126
thermodynamics, laws of, 146-149
Tulley, Brent, 118-119

universe age dilemma, 118-119
universe, distances in, 27-33
universe, mass density of,
   see, density of universe
universe, models of, 4-7
   big bang, 4-9, 36-38
      see also, big bang theory
   expansion is apparent, 5, 34-35
   steady-state universe, 5, 10-11, 35-36,
   Yilmaz cosmology model,
      see, Yilmaz cosmology model
universe radius,
   see, observable universe
universe, structure of, 118-119
   (filament and ribbon)
unified field theory, 91-93, 96, 106, 130-131

vector, 59-62
vertebrate, 16-17

wave, electromagnetic, 11, 66-68, 145
wave, gravitational, 11, 104, 145, 152
wave, light, 11, 65-66, 68-70, 74, 114
wave , radio, 11, 46, 68
wave, sound, 65-66, 68-69
wave, water, 65-66, 68
wavelength, 4, 23, 30, 39-42, 48, 70, 86-87, 90, 107, 114, 139, 160
Wilson, Robert, 38, 41

Yilmaz, Huseyin, 10, 11, 102-107, 109-110, 124, 126-127, [Y1 to Y12]

Yilmaz gravity theory, 10-13, 34, 42, 45, 49-50, 102-107, 109-114, 124, 126-127, 129-131, 144-146, 152

Yilmaz cosmology model, 10-13, 34, 42, 50, 129-131, 132-143, 144-146. 148-150, 152, 157-158, 160-161, 163

wavelength, 4, 23, 30, 39-42, 48, 70, 86-87, 90, 107, 114, 139, 160

Young, Thomas, 69